Fundamentals of Electromagnetics 1: Internal Behavior of Lumped Elements

Fundamentals of Electromagnetics 1: Internal Behavior of Lumped Elements

David Voltmer

ISBN: 978-3-031-79413-1   paperback
ISBN: 978-3-031-79414-8   ebook

DOI 10.1007/978-3-031-79414-8

A Publication in the Springer series
SYNTHESIS LECTURES ON COMPUTATIONAL ELECTROMAGNETICS #14

Lecture #14
Series Editor: Constantine A. Balanis, Arizona State University

Library of Congress Cataloging-in-Publication Data

Series ISSN: 1932-1252   print
Series ISSN: 1932-1716   electronic

First Edition
10 9 8 7 6 5 4 3 2 1

Fundamentals of Electromagnetics 1: Internal Behavior of Lumped Elements

David Voltmer
Rose-Hulman Institute of Technology

SYNTHESIS LECTURES ON COMPUTATIONAL ELECTROMAGNETICS #14

ABSTRACT

This book is the first of two volumes which have been created to provide an understanding of the basic principles and applications of electromagnetic fields for electrical engineering students. *Fundamentals of Electromagnetics Vol 1: Internal Behavior of Lumped Elements* focuses upon the DC and low-frequency behavior of electromagnetic fields within lumped elements. The properties of electromagnetic fields provide the basis for predicting the terminal characteristics of resistors, capacitors, and inductors. The properties of magnetic circuits are included as well. For slightly higher frequencies for which the lumped elements are a significant fraction of a wavelength in size the second volume of this set, *Fundamentals of Electromagnetics Vol 2: Quasistatics and Waves*, examines how the low-frequency models of lumped elements are modified to include parasitic elements. Upon completion of understanding the two volumes of this book, students will have gained the necessary knowledge to progress to advanced studies of electromagnetics.

KEYWORDS

Electromagnetics Introduction, Electromagnetic fields within lumped elements, Terminal characteristics of resistors, capacitors, and inductors, Properties of magnetic circuits

Contents

Preface

"Learning about 0's and 1's is much easier than learning about Electromagnetics."
—Robert MacIntosh

Most students begin their studies in electrical engineering with courses in computer logic and electric circuits. This is quite understandable since the students' mathematical background is usually limited to scalar mathematics. But, the fundamental principles of electrical engineering upon which the majority of physical processes such as semiconductor devices, power generation, or wireless technology are represented mathematically by three-dimensional, vector calculus. Electromagnetics is a very challenging subject for students lacking a strong background in mathematics or three-dimensional visualization skills—even for students who have a thorough understanding of electric circuits. On the other hand, the principles of electromagetics are relatively simple when the underlying vector calculus is understood by the students. This textbook is a non-traditional approach to bridge this gap that is based upon the similarities of lumped passive elements of circuits—resistors, capacitors, and inductors.

The terminal characteristics of lumped elements are derived from the observable quantities of voltage difference and current flow that are measured with instruments of finite size, i.e., dimensions on a macro-scale. But a more fundamental description of the underlying phenomena is on a micro-scale, infinitesimal in size, i.e., smaller than the smallest instrument. This behavior is described in terms of differential dimensions, with dimensions that shrink to zero; it is not measurable by man-made instruments and is much more complicated than two terminal properties of voltage drop and current flow. These inner workings of lumped elements on a micro-scale are related to their external, observable macro-scale (or terminal) characteristics.

This approach utilizes the following set of underlying restrictions:

- The terminal behavior of passive, lumped elements, $v_R = i_R R$, $i_C = C dv_C/dt$, and $v_L = L di_L/dt$, is known from circuits. Circuit theory experience, especially with KVL and KCL, is assumed and used.

- Each material type has only a single electromagnetic property—conductors have non-zero conductivity, σ, and allow current flow; dielectrics have permittivity, ε, and store only electric energy; and magnetic materials have permeability, μ, and store only magnetic energy.

- Lumped elements of finite size can be decomposed into discrete, curvilinear regions each of which behaves as an incremental lumped element. The total element value is calculated using the series/parallel equations of circuits.

- The electric or magnetic scalar potentials of all three types of lumped elements satisfies Laplace's equation, $\nabla^2 V = 0$. The same solution methods can be used for all elements.

- There are no fringing or leakage fields from any lumped elements. Consequently, all inductors have torroidal-shaped, magnetic cores.

- The frequency of operation is low enough that the elements are small compared to wavelength. Consequently, radiation and wave effects are ignored.

- Linearity and superposition apply with one notable exception—magnetic circuits.

- All electrodes and connecting wires are lossless with zero resistance and as such are called Perfect Electric Conductors (PEC).

These restrictions are imposed so that the basic principles of electromagnetics are not hidden by complicating geometric or mathematical details. Those details can come in following courses. Those students who take more advanced electromagnetic courses are not hindered by this approach. In fact, a case may be made that this approach provides a more complete understanding of the basics than traditional approaches.

Analytic expressions are used throughout this text as a calculation technique. Numeric and graphical methods for two dimensional structures are integrated within the textbook as well. In addition, circuit simulation is introduced as a viable solution method.

Finally, It is assumed that all students have studied electricity and magnetism in introductory physics and have an understanding of the concepts of force, displacement, work, potential energy, opposite charges attract, like charges repel, and current is usually assumed to be uniformly distributed currents in resistors. In addition, students should be familiar with vector algebra. Most work uses Cartesian coordinates with a few examples in cylindrical and spherical coordinates. The SI system of units is used throughout.

Dr. David R. Voltmer   May 2007
Professional Engineer
Professor Emeritus
Electrical and Computer Engineering
Rose-Hulman Institute of Technology

CHAPTER 1

Resistors

1.1   RESISTORS: A FIRST GLANCE

Resistors are the simplest of lumped elements that you already know from your circuits courses. They enable us to control voltage drops and current flow throughout lumped element circuits. The application of a voltage drop across a resistor causes a current to flow into the more positive terminal, through the resistor, and out of the negative terminal. The relationship of the voltage drop across a resistor, V_R, to the current through it, I_R, is known as Ohm's law and is expressed as

$$R = \frac{V_R}{I_R} \ [\Omega] \qquad (1.1)$$

where R is the value of the resistance expressed in ohms and denoted by the symbol Ω. The resistance value is a function of the geometry and material composition of the resistor. This chapter focuses upon the internal workings of the resistor and how to calculate its resistance.

The structure of a typical resistor is of the form sketched in Fig. 1.1. The wire leads and the electrodes of a resistor are made of a good conductor, usually copper, which has a very small resistance. That means that there is a very little voltage drop across the electrodes regardless of the magnitude of the current. Current flow between the two electrodes is through the conductive material between them.

The conductive material greatly impedes the current and governs the resistor's value. The conductive material of resistors can take on a wide variety of composition and shape depending upon the application for which the resistor is intended. This structure is usually hermetically encapsulated to protect it against external substances that could alter its value.

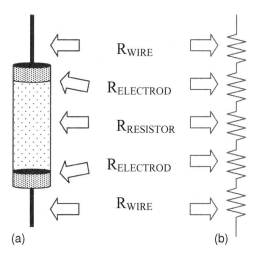

FIGURE 1.1: Structure of a resistor: (a) physical view and (b) electrical model.

1.2   MODELING AND APPROXIMATIONS

Before proceeding further in the analysis of resistors, let's be clear about our strategy in solving electromagnetic problems. In order to understand the electrical behavior of a resistor, we must first devise a model which retains the resistor's essential electrical characteristics. There is no unique model; different solution methods or different features to be studied will lead to different models—the simpler the model, the easier the analysis. But the model must retain the essential features which produce the behavior to be studied. Our engineering judgment is needed to make the tradeoff between simplicity and accuracy in our model.

To gain an understanding of electromagnetic phenomena, we do not need solution accuracy to 0.01%. This means that a little approximation here and there will be acceptable so long as it makes a problem solvable. Approximations may simplify our model, allowing a simple solution. After we obtain a solution, we must check the results with our approximations to make sure that they are still valid. If so, then we have been successful; if not, we must rethink the problem and try another model. Engineering experience, logical thinking, and common sense are an integral part of our efforts in modeling and approximation. Good engineers use this strategy over and over throughout their careers—master it!

Now, let's return to the structure of a resistor. Current in the resistor flows through the metallic leads, the electrodes, and the conductive material. Each of these components can be modeled as an ideal resistor. Since the same current flows through each of these resistors, we connect them in series. This model of a resistor is shown in Fig. 1.1(b). The metallic leads and the electrodes have a very small resistance ($R_{\mathrm{WIRE}} + R_{\mathrm{ELECTRODE}} \ll 1\,\Omega$) which is negligible compared to the resistance value of most resistors. Therefore, the wire and electrodes may be approximated as zero resistance, $R_{\mathrm{WIRE}} + R_{\mathrm{ELECTRODE}} = 0\,\Omega$. Consequently, the entire voltage drop across the resistor is approximated as contained entirely within the conductive material. The value of the resistor is determined only by the composition and configuration of the conductive material between the electrodes.

1.3   VOLTAGE DROP, VOLTAGE, AND FIELDS

You recall from elementary physics that the resistance of a long cylindrical wire is given by

$$R = \frac{L}{A\sigma} \tag{1.2}$$

where L is the length of the resistor, A is the cross-sectional area, and σ is the conductivity of the wire. This equation refers to a solid cylinder of conductive material with its length measured perpendicular to the constant cross-sectional area. Such a resistor is shown in the circuit of Fig. 1.2. A DC or slowly varying voltage source is applied to the resistor. With an

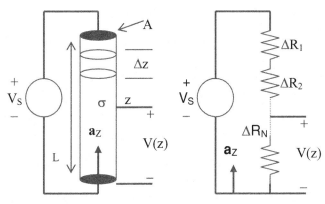

FIGURE 1.2: Resistor circuit and model.

ideal voltmeter, we can measure the voltage drop between any two points along the axis of the resistor, $V(z)$.

Each incremental segment Δz acts as an incremental resistor with a resistance calculated from Eq. (1.2) as

$$\Delta R = \frac{\Delta z}{A\sigma}. \tag{1.3}$$

Since the current in each of the incremental resistors is the same, the total resistance is modeled as a series connection of incremental resistors and is equal to the sum of the incremental resistances as

$$R = \sum_{i=1}^{N} \Delta R_i = \sum_{i=1}^{N} \frac{\Delta z_i}{A\sigma} = \left(\frac{1}{A\sigma}\right) \sum_{i=1}^{N} \Delta z_i. \tag{1.4}$$

Note that the cross-sectional area A and the conductivity of the material σ have been assumed to be the same for all values of z and, thus, factored out of the summation. When the incremental length is replaced by a differential length, the summation becomes an integral and Eq. (1.4) becomes

$$R = \left(\frac{1}{A\sigma}\right) \int_{z=0}^{L} dz. \tag{1.5}$$

Note that a coordinate system must be defined to use the integral representation. In Fig. 1.2, the axis of the resistor is aligned conveniently with the z-axis which points upward; the bottom of the resistor is at $z = 0$ and the top at $z = L$. The resistance $R(z)$ measured from position z to ground is given by

$$R(z) = \left(\frac{1}{A\sigma}\right) \int\limits_{z'=0}^{z} dz' = \frac{z}{A\sigma}. \tag{1.6}$$

If we were to measure the voltage drop across the resistor from a position z to the bottom end, we would find that it obeys the voltage divider equation as

$$V(z) - V(0) = V_S \frac{R(z)}{R(L)} = V_S \frac{z}{L} \tag{1.7}$$

where the top end is the more positive terminal. As you recall from circuits, when the voltmeter is connected to two different points in a circuit, the difference in voltage or voltage drop between those two points is measured. Often one of the leads is connected to a ground or reference node. All other nodal voltages are defined as the voltage difference between the node voltage and the reference voltage. In electromagnetics, voltage differences are measured or defined with respect to a convenient reference voltage. The zero potential may be defined as at a ground plane, at a negative of a power supply, or at a distant point infinitely far away. All voltage drops are defined or measured relative to this location. Moreover, usually it is just called a *voltage rather than a voltage drop*.

For the resistor shown in Fig. 1.2, we can define the lower end of the resistor as ground or zero volts. With this understanding, we can rewrite Eq. (1.7) for the voltage at some point z in the resistor as

$$V(z) = V_S \frac{R(z)}{R(L)} = V_S \frac{z}{L}. \tag{1.8}$$

When written in this form, there is a functional relationship between every point along the axis of the resistor and the voltage at that point. Mathematicians define such a functional dependence as a field; the voltage is a field over the spatial variable z. This is a scalar field since at each point z there is a single number which represents the voltage $V(z)$.

Example 1.3-1.   Calculate the resistance of the copper wire leads of a resistor. Estimate each lead as $1''$ long and 1 mm diameter. From Appendix D, we find that $\sigma_{COPPER} = 5.8 \times 10^7 \ \Omega^{-1} \ m^{-1}$. Using Eq. (1.2), we obtain $R_{LEAD} = \frac{L}{\sigma A} = \frac{2(0.0254)}{5.8 \times 10^7 \pi (0.0005)^2} = 1.1 \ m\Omega$. This is certainly much less than 1 Ω and verifies our earlier assumption.

Example 1.3-2.   Estimate the resistance of the lead in a pencil measured from one end of a pencil to the other. The lead is in the form of a circular cylinder of 2 mm diameter and 20 cm (about $8''$) length. Assume the lead is made of graphite. From Appendix D, we find that

$\sigma_{\text{GRAPHITE}} = 7 \times 10^4 \ \Omega^{-1} \ \text{m}^{-1}$. So, from Eq. (1.2), we obtain $R = \frac{L}{\sigma A} = \frac{0.2}{7 \times 10^4 \pi (0.001)^2} = 0.909 \ \Omega$.

Example 1.3-3. A toaster is designed for household use of 110 VAC (RMS). The heating element is made of a nichrome wire of 1 mm diameter. What length of the wire should be chosen to limit the current to 5 A? From Appendix D, we find that $\sigma_{\text{NICHROME}} = 10^6$ S/m. The required resistance is $R \geq 110/5 = 22 \ \Omega$. From Eq. (1.2), we obtain $L = R\sigma A = 22(10^6)\pi(0.0005)^2 = 17.3$ m.

1.4 ELECTRIC FIELD INTENSITY

Equation (1.8) reveals a very interesting feature of the voltage—the voltage drop $V(z)$ is proportional to the length of resistor from the ground. A plot of the $V(z)$ versus z is linear so that the voltage drop divided by the resistor length results in a constant, $V(z)/z = V_S/L$, the slope of the straight line. Extending this concept to cases of arbitrary voltage distributions with varying slope, we can approximate the slope at point z as

$$\frac{\Delta V(z)}{\Delta z} = \frac{V_S}{L} \tag{1.9}$$

with units of volts/meter as shown in Fig. 1.3. This incremental form is more revealing than the voltage difference between two points since it describes the incremental behavior of the voltage at each point z.

The property of the voltage expressed by Eq. (1.9) is so fundamental that it has been defined as the magnitude of a vector field parameter **E**, the *electric field intensity*. The vector **E** is defined to point in the direction from more positive voltage to less positive voltage. When

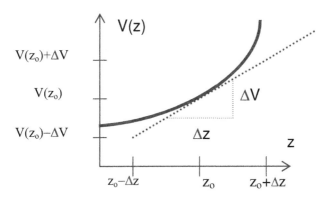

FIGURE 1.3: Voltage distribution $V(z)$ and slope $\Delta V/\Delta z$.

$\Delta V/\Delta z > 0$, $V(z)$ is increasing with increasing z so that **E** points in the direction of $-\mathbf{a}_Z$; when $\Delta V/\Delta z < 0$, $V(z)$ is decreasing with increasing z so that **E** points in the direction of $\mathbf{a}_Z$. Combining these concepts, we are led to a mathematical definition of the incremental electric field intensity as

$$\mathbf{E}(z) = -\frac{\Delta V(z)}{\Delta z}\mathbf{a}_Z. \qquad (1.10)$$

$\mathbf{E}(z)$ points in the direction of decreasing voltage and has a magnitude which is equal to the incremental changes in $V(z)$ with respect to incremental changes in z. The greater the spatial rate of change of $V(z)$ the greater $\mathbf{E}(z)$. Obviously, $\mathbf{E}(z)$ has units of volts/meter. Finally, note that $\mathbf{E}(z)$ is a vector field over the variable z, i.e., for every value of z there is a corresponding vector of value $\mathbf{E}(z)$. This is a vector field because it has a direction as well as a magnitude at each point z.

For the case of the resistor of Fig. 1.2, the electric field vector has only single component in the z-direction since that is the direction in which the voltage changes with respect to z. This z-component of the electric field intensity vector is expressed as

$$E_Z = -\frac{\Delta V(z)}{\Delta z}\,[\mathrm{V/m}]. \qquad (1.11)$$

An interpretation of the electric field intensity comes via the analogy of the voltage, $V(z)$ (measured with respect to a reference voltage), to the height of a point on a hill measured from its base or height reference (such as sea level). To move up the hill, we must move upward against the force of gravity; the steeper the hill, the greater the force. The magnitude of the force we must exert for each step, i.e., the force/meter, can be compared with the electric field intensity, **E**. The gravitational force points down the hill from higher to lower levels just as **E** points from higher to lower voltages. We will return to this analogy later.

Example 1.4-1. A $1\,\mathrm{M}\Omega$ resistor which measures $1/8''$ diameter and $1/2''$ long has a voltage drop of $100\,\mathrm{V}$ across it. Calculate the magnitude of the electric field within the resistor. From Eqs. (1.9) and (1.10), we find $|\mathbf{E}| = \Delta V/\Delta z = 100/(0.5)(0.0254) = 7.87\,\mathrm{kV/m}$.

1.5   GENERALIZED COORDINATES

But not all resistors lie along the z-axis. More generally, they need not have a straight-line axis. If the concept of electric field intensity is to be useful, it must work in more general situations. Let's consider a linear resistor which lies along some arbitrary straight line with one end at (x_1, y_1, z_1), the other at (x_2, y_2, z_2), and with a voltage drop from one end to the other as shown in Fig. 1.4, where P2 is the more positive terminal and P1 is the less positive terminal. The projection of the resistor length onto the x-axis is $L_X = (x_2 - x_1)$; L_Y and L_Z are defined

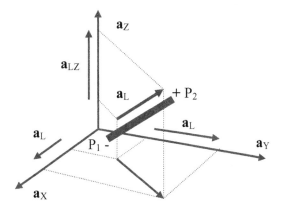

FIGURE 1.4: Arbitrarily-oriented resistor.

similarly. The length of the resistor is given by $L = [L_X^2 + L_Y^2 + L_Z^2]^{1/2}$. Of course, we know that the linear variation of the voltage from one end of the resistor to the other will not be altered by its spatial orientation.

Consequently, $\mathbf{E}$ is directed along the axis of the resistor from the more positive end to the less positive end. As before, $|\mathbf{E}|$ is equal to $|\Delta V/\Delta l|$ where l is distance measured along the axis of the resistor. This is expressed as

$$\mathbf{E} = -\frac{V_S}{L}\mathbf{a}_L \qquad (1.12)$$

where $\mathbf{a}_L$ is directed from P_1 to P_2. $\mathbf{E}$ can be decomposed into rectangular components as

$$\mathbf{E} = E_X\mathbf{a}_X + E_Y\mathbf{a}_Y + E_Z\mathbf{a}_Z. \qquad (1.13)$$

Each of the components is determined as the projection of $\mathbf{E}$ on its coordinate axis. As you recall from vector algebra (see Appendix B), the ith component, E_i, can be found from the dot product as

$$E_i = \mathbf{a}_i \cdot \mathbf{E} = |\mathbf{E}| \cos \gamma_i \qquad (1.14)$$

where γ_i is the angle between the axis of the resistor, $\mathbf{a}_L$, and the coordinate axis, $\mathbf{a}_i$, e.g., $\cos \gamma_X = L_X/L$. For a general voltage distribution where V is a function of three spatial variables, i.e., $V = V(x, y, z)$, each component E_i represents the incremental change of voltage due to an incremental change of position in the ith coordinate direction. Recall that a change in the voltage due to a change in position Δz defines the component of the electric field intensity in the z-direction, $E_z = -\Delta V/\Delta z$. Extending this concept, we define the other two components as $E_X = -\Delta V/\Delta x$ and $E_Y = -\Delta V/\Delta y$. Substituting these forms into Eq. (1.13) leads to

$$\mathbf{E} = -\left(\frac{\Delta V}{\Delta x}\mathbf{a}_X + \frac{\Delta V}{\Delta y}\mathbf{a}_Y + \frac{\Delta V}{\Delta z}\mathbf{a}_Z\right). \qquad (1.15)$$

Each component of the electric field intensity is associated with the change in voltage with respect to the change in position in the particular direction. Since most material is linear with respect to electromagnetic fields, the individual vector components of **E** are added to obtain the total electric field intensity. This means that **E** points in the direction of the greatest rate of change of voltage with respect to distance. When the voltage changes in a particular direction are large, that component of the electric field is large and **E** points mainly in that direction. Of course, each of the components is less than the magnitude of the field, i.e., $E_i \leq |\mathbf{E}|$. If we relate the height of a hill to the voltage, then the force one feels while on the hill compares with **E** in a two-dimensional voltage distribution. Just as the force points down the hill in the direction of the steepest slope, **E** points in the direction of greatest decrease in voltage.

In summary, the electric field intensity is governed by the spatial variations of the voltage. The electric field points in the direction of greatest negative spatial rate of change of the voltage with a magnitude equal to the rate of change of voltage in that direction.

Example 1.5-1.   Consider that the axis of the resistor of Example 1.4-1 lies on the line from the origin, $P_1 = (0, 0, 0)$ to $P_2 = (1, 1, 2.45)$. Furthermore, connect the origin as the negative end of the resistor. Calculate **E**, E_X, E_Y, and E_Z. From Example 1.4-1, we already know $|\mathbf{E}| = 7.87$ kV/m. To find the orientation of the line, we use the relations for the spherical angles from Appendix B as $\theta = \arccos(z/[x^2 + y^2 + z^2]^{1\backslash 2}) = 30°$ and $\phi = \arccos(x/[x^2 + y^2]^{1/2}) = 45°$. Therefore, $\cos \gamma_X = \sin \theta \cos \phi = 0.354$, $\cos \gamma_Y = \sin \theta \sin \phi = 0.354$, and $\cos \gamma_Z = \cos \theta = 0.866$. (Note that these results appear to be correct since the sum of their squares is 1 as it should be.) Components are calculated from Eq. (1.14) as $E_X = E_Y = (0.354)(7.87 \text{ kV/m}) = 2.79$ kV/m and $E_Z = (0.866)(7.87 \text{ kV/m}) = 6.82$ kV/m. Since the point (1, 1, 2.45) is the more positive terminal, **E** points back toward the origin so that all the components will have a negative sign. Therefore, $E_X = E_Y = -2.79$ kV/m and $E_Z = -6.82$ kV/m and $\mathbf{E} = -2.79\mathbf{a}_X - 2.79\mathbf{a}_Y - 6.82\mathbf{a}_Z$ kV/m.

1.6   VOLTAGE GRADIENT

Many resistors do not conform to the simple cylindrical nature we have just considered. Rather they come in a variety of shapes with many different electrode configurations. They posses a general voltage distribution $V(x, y, z)$ for which the electric field is given by Eq. (1.15). The right-hand side of this equation has a form that occurs in all areas of engineering and physics. Accordingly, mathematicians have devoted great effort to the rigorous development of this mathematical operation of vector calculus—derivatives and integrals of vector fields. If we consider Eq. (1.15) in the limit as the incremental displacements of position approach zero, then we arrive at the definition of this mathematical operation, known as the *gradient*. It is expressed as

$$\nabla V = \lim_{\Delta x, \Delta y, \Delta z \to 0} \left(\frac{\Delta V}{\Delta x} \mathbf{a}_X + \frac{\Delta V}{\Delta y} \mathbf{a}_Y + \frac{\Delta V}{\Delta z} \mathbf{a}_Z \right)$$

$$= \frac{\partial V}{\partial x} \mathbf{a}_X + \frac{\partial V}{\partial y} \mathbf{a}_Y + \frac{\partial V}{\partial z} \mathbf{a}_Z. \tag{1.16}$$

The nabla or del operator, ∇, defines a series of derivative operations on a scalar function that produces a vector result. In some literature, the gradient of V is called grad(V). The behavior of V at a point is defined by this operation. In heat transfer, the flow of heat energy is in the direction of and proportional to the magnitude of $-\nabla T$ where T is the temperature as a function of position. In fluids, the flow of material is in the direction of and proportional to the magnitude of $-\nabla p$ where p is the pressure within the fluid as a function of position. In all cases, the vector direction is from more positive values of the scalar to less positive values as it is with $-\nabla V$ and V.

Throughout this textbook, we will emphasize rectangular coordinates as we have done in the development of the gradient. However, all of these concepts are valid in many other coordinate systems. The gradient operators in cylindrical and spherical coordinates are given in Appendix C.

The use of differentials in the mathematical description of the gradient in Eq. (1.16) is useful for analytic solutions. However, for numeric work, the incremental or difference form of the gradient in Eq. (1.15) is more useful. Of course, in the limit as the increments approach zero, the two forms are the same.

The expression for $\mathbf{E}$ of Eq. (1.15) can be expressed in terms of the gradient as

$$\mathbf{E} = -\nabla V \tag{1.17}$$

so long as the derivatives required in the gradient exist. This is a useful mathematical concept, but it is really not observable, not measurable, since it is defined at a single point and all of our measuring equipment occupies a finite region of space. Nevertheless, it is a useful concept that we can always approximate by Eq. (1.15) which can be measured.

Example 1.6-1. Calculate $\mathbf{E}$ for the voltage distribution of $V(x, y, z) = 4x \cos y - yz$ V and evaluate it at $(1, -\pi/4, -2)$. Implementing Eq. (1.17), we find $\mathbf{E} = -\nabla V = -[4 \cos y \mathbf{a}_X - (4x \sin y + z) \mathbf{a}_Y - y \mathbf{a}_Z]$ so that $\mathbf{E}(1, \pi/4, -2) = -2.82 \mathbf{a}_X + 0.828 \mathbf{a}_Y + 0.785 \mathbf{a}_Z$ V/m.

1.7   ∇V AND EQUIPOTENTIALS

As we saw in the last section, ∇V represents the magnitude and direction of the greatest spatial rate of change of voltage. The spatial rate of change of voltage in any other direction is less than this maximum value. The component of $\mathbf{E}$ in a particular direction is described by Eq. (1.14) as

$$\mathbf{E}_i = \mathbf{E} \cdot \mathbf{a}_i = -\nabla V \cdot \mathbf{a}_i = -\frac{\partial V}{\partial l_i} \qquad (1.18)$$

where $\mathbf{a}_i$ is the unit vector and l_i is the distance in the specified direction. The component of the electric field in the ith direction is the spatial rate of change of the voltage in that direction. Mathematicians have given this the name of *directional derivative* since it represents the rate of change of the voltage in a particular direction. For finite differences, it represents the incremental change of voltage, ΔV_i, with respect to distance measured in the ith direction, Δl_i.

When $\mathbf{a}_i$ is perpendicular to $\mathbf{E}$, the dot product of Eq. (1.18) becomes zero indicating that there is no change in voltage for displacements in that direction. Since there is no change in voltage, i.e., $\Delta V_i = 0$, for an incremental displacement perpendicular to $\mathbf{E}$, the potential remains constant along this displacement. There are an infinite number of directions perpendicular to $\mathbf{E}$; they form a surface to which $\mathbf{E}$ is perpendicular. Since the potential on this surface is unchanging it is called an *equipotential surface* and is illustrated in Fig. 1.5. Note that what we call voltage is also called potential; hence, these equivoltage surfaces are called equipotential surfaces.

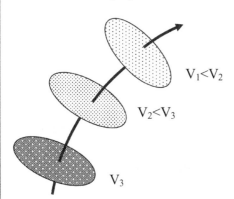

$V_1 < V_2$

$V_2 < V_3$

V_3

FIGURE 1.5: $\mathbf{E}$ and several equipotential surfaces.

The electric field intensity $\mathbf{E}$ is always perpendicular to equipotential surfaces! This follows from the fact that $\mathbf{E}$ is in the direction of the greatest change of voltage whereas an equipotential surface has no potential change.

Returning to the analogy of the height of a hill to a two-dimensional voltage distribution, we see that horizontal contours on a topographical map of a hill's surface are analogous to equipotential lines or contours in a two-dimensional voltage field. The force down the hill is perpendicular to the contour since $\mathbf{E}$ is perpendicular to equipotential surfaces or contours.

But there is an additional concept of importance here. Since $\mathbf{E} = -\nabla V$, then ∇V is also perpendicular to equipotential surfaces. Since the perpendicular or normal to a surface is not uniquely defined, the opposite direction, $-\nabla V$, is also perpendicular. This property can be used to find the vector perpendicular to a surface, i.e., the surface normal, in the following manner. For an equipotential surface defined by $V(x, y, z) = V_o$, the unit vector normal to the surface (pointing in the direction of increasing V) is given by

$$\mathbf{a}_N = \frac{\nabla V}{|\nabla V|}. \tag{1.19}$$

Note that since the potential $V(x, y, z)$ is a spatial function, the unit surface normal $\mathbf{a}_N$ varies from point to point as well. The unit vector is calculated by dividing the surface normal ∇V by its length $|\nabla V|$.

Example 1.7-1.   Find the normal to the equipotential surface of Example 1.6-1 at the point $(1, \pi/4, -2)$. We already have $-\nabla V = -2.828\mathbf{a}_X + 0.828\mathbf{a}_Y + 0.785\mathbf{a}_Z$ at this point so that $|\nabla V| = [(2.828)^2 + (0.828)^2 + (0.785)^2]^{1/2} = 3.05$ and $\mathbf{a}_N = \pm\frac{\nabla V}{|\nabla V|} = \pm(-0.927\mathbf{a}_X + 0.271\mathbf{a}_Y + 0.257\mathbf{a}_Z)$. Note the ambiguity in direction for the normal since there is no guide as to which direction is defined as the positive normal.

1.8   VOLTAGE DROP AND LINE INTEGRALS

The work to this point provides a basis for expressing the microscopic behavior of the voltage and the related electric field intensity within a resistor. But this does not give a measurable voltage drop between two points of a resistor. $V(x, y, z)$ is the voltage relative to the voltage at some reference location. On the other hand, the measurement of voltage drop gives the difference between the voltages at two distinct points—the voltage of one point *relative* to another, not the reference voltage. In this section, we will devise a way to express this relative difference.

Recall that a component of the electric field intensity can be expressed as

$$E_i = -\frac{\Delta V_i}{\Delta l_i} \tag{1.20}$$

where E_i is the component of $\mathbf{E}$ in the direction in which Δl_i is measured.

The incremental voltage difference ΔV_i between the two endpoints of the incremental distance Δl_i, is given as

$$\Delta V_i = -E_i \Delta \ell_i = -\mathbf{E} \cdot \mathbf{a}_i \Delta \ell_i \tag{1.21}$$

using the well-known relationship $E_i = \mathbf{E} \cdot \mathbf{a}_i$. The last two terms can be combined as $\mathbf{a}_i \Delta l_i = \Delta l_i$ and interpreted as a directed incremental line segment. This general form can be used to calculate the incremental voltage difference of all incremental line segments. All curved lines,

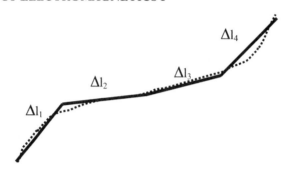

FIGURE 1.6: Smooth path and segmental approximation.

more commonly called paths, can be approximated as a sequence of incremental line segments as shown in Fig. 1.6.

Consequently, the voltage difference between the endpoints of the path can be obtained as the sum of the incremental voltage differences as

$$V_{\text{FINAL}} - V_{\text{INITIAL}} = \sum_{i=1}^{N} \Delta V_i = -\sum_{i=1}^{N} \mathbf{E}_i \cdot \mathbf{a}_i \Delta \ell_i$$

$$= -\sum_{i=1}^{N} \mathbf{E}_i \cdot \Delta \mathbf{l}_i. \tag{1.22}$$

The directed line segment $\Delta \mathbf{l}_i$ approximates the tangent to the smooth path in the ith segment. Therefore, the dot product $E_i \cdot \Delta \mathbf{l}_i$ depends upon the orientation of the electric field intensity and the directed path at each point as does the total voltage difference. The smaller the increment, $\Delta \mathbf{l}_i$, the more accurately the line segments approximate the smooth path and the more accurately is the voltage difference calculated. As the incremental distance approaches zero, the sum can be replaced by an integral form

$$V_{\text{FINAL}} - V_{\text{INITIAL}} = -\int_{L} \mathbf{E} \cdot \mathbf{dl} \tag{1.23}$$

where L represents the path. This integral is known to mathematicians as a *line integral*. When the path L extends from one electrode of a resistor to the other, Eq. (1.23) represents the voltage drop across the resistor. This is the voltage we would expect to measure with a voltmeter.

Often the path $L = L(x, y, z)$ and the differential $\mathbf{dl}$ can be expressed as a function of position. This means that the integral depends upon the path so that two different paths through the resistor from one electrode to the other could produce different voltage drops. But, two different voltage drops is contrary to Kirchoff's voltage law (KVL) of circuit theory and requires a closer look.

1.9 KVL AND CONSERVATIVE FIELDS

One of the foundations of circuit theory is Kirchoff's voltage law which states that the sum of the voltage drops around a closed loop is zero. Alternatively, this means that for any two paths of a circuit from point B to point A the voltage drop must be same, i.e., $V_{AB1} = V_{AB2}$, see Fig. 1.7. Note the conventional circuit formulation for voltage drop has been used here as $V_A - V_B = V_{Final} - V_{Initial} = V_{AB}$.

If the field description of a resistor is to be consistent with circuit theory, KVL must hold, i.e., the sum of the voltage drops around a closed path must be zero. In integral form, this can be stated as

$$V_{AB1} = -\int_{L_1} \mathbf{E} \cdot \mathbf{dl} = V_{AB2} = -\int_{L_2} \mathbf{E} \cdot \mathbf{dl}; \qquad (1.24)$$

correspondingly, the voltage around the closed path is

$$V_{AB1} - V_{AB2} = -\int_{L_1} \mathbf{E} \cdot \mathbf{dl} + \int_{L_2} \mathbf{E} \cdot \mathbf{dl}$$

$$= -\int_{L_1 - L_2} \mathbf{E} \cdot \mathbf{dl} = -\oint_L \mathbf{E} \cdot \mathbf{dl} = 0. \qquad (1.25)$$

The integral sign with the closed circle superimposed upon it, $\oint_L$, is the mathematicians shorthand for indicating the integral is evaluated on the closed path L; the path begins and ends at the same point. In Fig. 1.7, the closed path L is composed of the integral on L_1 in the positive direction followed by the integral on L_2 in the negative direction. For incremental line segment approximations of the path L, this is expressed as

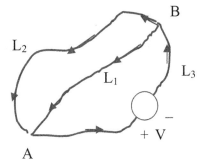

$$\sum_{i=1}^{N} \mathbf{E}_i \cdot \Delta \mathbf{l}_i = 0. \qquad (1.26)$$

FIGURE 1.7: Application of Kirchoff's voltage law.

But, what is the meaning of these forms? Consider KVL applied to the closed loop of Fig. 1.7 that includes the ideal voltmeter, V. Note that an ideal voltmeter has infinite impedance and has not current flow. The leads attached to the voltmeter are assumed to have no resistance so they have no voltage drop regardless of the current flow. The voltage drops along the directed path L_1 occurs across the resistor, V_{AB1}, and along the directed path L_3 across the voltmeter, V_{METER} (no voltage drop in the leads of an ideal voltmeter), with the polarity shown; they add to zero since

$$V_{AB1} - V_{METER} = - \oint_{L_3} \mathbf{E} \cdot \mathbf{dl} = 0. \qquad (1.27)$$

The voltmeter will indicate the voltage drop across the resistor, i.e., $V_{AB1} = V_{METER}$, as we know from circuits laboratory. The path of the integral has been chosen to be coincident with the branches of the circuit in Fig. 1.7. Similarly, using path L_2 instead of L_1 gives the result $V_{AB2} = V_{METER}$. More general applications for paths that are not aligned with the branches of a circuit will be considered later.

An alternate perspective of this concept is gained by considering the effects of $\mathbf{E}$ on a charged particle Q as we move it along a curved path. When a stationary charged particle Q is present in an electric field, it experiences a force which has been experimentally measured to be

$$\mathbf{F}_{ELECT} = Q\mathbf{E}. \qquad (1.28)$$

This is called the Coulomb force in honor of the experimental work of Colonel Charles Coulomb in establishing Eq. (1.28). Note that the force on the charge and the electric field are in the same direction for a positive charge. To keep the charge from being accelerated by this force (remember $\mathbf{F} = m\mathbf{a}$), a mechanical force must be applied to the charge which is equal and opposite to the force caused by the electric field, i.e., $\mathbf{F}_{MECH} = -\mathbf{F}_{ELECT}$. When this mechanical force is greater than the electric force, the charge is accelerated in the direction opposite the electric field. Imagine applying to the charge a mechanical force that is ever so slightly greater than the electrical force, moving the charge very slowly so that effects of acceleration are negligible. The mechanical force does work on the charge according to the well-known principle of mechanics

$$\Delta W_{MECH} = \mathbf{F}_{MECH} \cdot \Delta \mathbf{l} = -Q\mathbf{E} \cdot \Delta \mathbf{l}. \qquad (1.29)$$

This is just Eq. (1.22) multiplied by the charge, Q; the incremental work done, ΔW_{MECH}, is equal to the charge multiplied by the incremental change in voltage, $Q\Delta V$. This provides an alternative definition of *voltage drop as the work done per unit charge* in moving a charge within an electric field,

$$\Delta V = \frac{\Delta W_{MECH}}{Q} = -\mathbf{E} \cdot \Delta \mathbf{l}. \qquad (1.30)$$

Recall from physics that the charged particle Q has an increased potential energy as a result of the work done, ΔW_{MECH}. The increase in potential energy per unit charge, $\Delta W_{MECH}/Q$, often is called the increase in potential. This is where the name potential came into use as an alternative for voltage when working with electromagnetic fields.

The mechanical work done as expressed by Eq. (1.30) depends upon the path chosen. The potential energy gained, i.e., the work done, in moving on a curved path L, has the same mathematical form as voltage drop

$$W_{\text{FINAL}} - W_{\text{INITIAL}} = \sum_{i=1}^{N} \Delta W_i = -Q \sum_{i=1}^{N} \mathbf{E}_i \cdot \Delta \mathbf{l}_i$$

$$= -Q \oint_L \mathbf{E} \cdot \mathbf{dl} = Q(V_{\text{FINAL}} - V_{\text{INITIAL}}). \qquad (1.31)$$

In the mechanical domain, we know that if an object is moved in a closed path on a frictionless surface, the total work done is zero. Equation (1.31) shows this as well since the mechanical work done on the left-hand side is zero when returning to the starting point. This corresponds to the change in voltage on the right-hand side is zero for a closed path, an expression of KVL.

Equation (1.31) also strengthens the analogy between the height of a hill and the voltage within a resistor. The mechanical work required to move up a hill results in the positive work done on the object just as work is required to move a charged particle up a potential hill. The mechanical work done is greatest when moving up the hill in the direction of the steepest slope; this is analogous to moving a charged particle in the direction opposite to the electric field. No mechanical work is done moving on a contour of constant height since there is no force of gravity in the direction of motion, i.e., the dot product between force and motion is zero since they are oriented 90° apart. Similarly, when a charge is moved along an equipotential, no work is done. The voltage drop measured perpendicular to an electric field is zero since this is along an equipotential.

Line integrals are also known as *work integrals* since they express the relationship of a vector field and the work done by moving along a prescribed path. Any field for which the work integral around a closed path is zero is known as a *conservative field*. Energy is conserved while traversing any closed path. Mathematicians have shown that any field which is conservative, i.e., has a work integral that is zero for every closed path, can be expressed as the gradient of a scalar. The electric field intensity in a resistor and the gravity force field are examples of conservative fields. They can be expressed as the gradient of a scalar quantity known as voltage or potential. But, beware, not all fields are conservative. In fact, time-varying electric fields when not confined to a relatively small region of space such as in a resistor are nonconservative. This means that later in our studies, we will have to generalize our description of an electric field. But, the cost of this extra work will be worth it, since the nature of nonconservative electric fields is what enables wave propagation to take place. For the time being, however, we will be content to consider conservative versions of electric fields only.

1.10 EVALUATION OF LINE INTEGRALS

Line or work integrals are the basis for many important calculations in electromagnetics. They will serve you well only if you master the techniques for their evaluation. This section provides the details for you.

Line integrals are defined by Eq. (1.23) as

$$V_{\text{FINAL}} - V_{\text{INITIAL}} = -\int_L \mathbf{E} \cdot \mathbf{dl}. \qquad (1.32)$$

The evaluation strategy is to first form the dot product between the vector integrand and the vector differential element. The functional description of the path $L(x, y, z)$ is then incorporated into the integrand and the differential element, giving them a representation unique to the path of integration. Straightforward integration of the several resultant scalar integrals completes the process. Let's look at the steps in more detail. For convenience, this discussion will focus upon Cartesian coordinates, but the procedure is similar for cylindrical and spherical coordinates.

Step 1: Form the Dot Product of the Integrand and the Differential. A vector field can be expressed in terms of its Cartesian components, $\mathbf{E} = E_X \mathbf{a}_X + E_Y \mathbf{a}_Y + E_Z \mathbf{a}_Z$. The differential is a vector as well and in general is described as $\mathbf{dl} = \mathbf{a}_X dx + \mathbf{a}_Y dy + \mathbf{a}_Z dz$. There is no direction or sign attached to the three scalar differentials. Instead, the direction is governed by our choice of endpoints for the integration path. The dot product within the integral becomes $\mathbf{E} \cdot \mathbf{dl} = E_X dx + E_Y dy + E_Z dz$.

Forming the dot product is really easy! In general, each of the components is a function of position, i.e., $E_i = E_i(x, y, z)$. Now we need to establish the relationship between the variables (x, y, z) which exist on the path of integration.

Step 2: Incorporate the Effects of the Path on the Integrand. From analytic geometry, we know that a surface can be described by a single equation among the variables, a line by two equations, and a point by three equations. Consequently, it takes at least two equations to represent the integration path. Often, these equations take the form of a relationship between two variables, e.g., $y = f(x)$ and $z = g(x)$, so that the functional dependence of a component E_i on the path of integration can be expressed in terms of a single variable, e.g., $E_Y(x, y, z) = E_Y(x, f(x), g(x))$.

The differentials of the two equations that define the path provide equations relating the scalar differentials, e.g., $dy = [\partial f(x)/\partial x]dx$ and $dz = [\partial g(x)/\partial x]dx$. Combining the vector components with the differentials for the example above, we obtain the three dot products as $E_X dx$, $E_Y dy = E_Y(x, f(x), g(x))[\partial f(x)/\partial x]dx$, and $E_Z dz = E_Z(x, f(x), g(x))[\partial g(x)/\partial x]dx$. Sometimes it is more convenient to write $x = f^{-1}(y)$ and express dx in terms of dy as $dx = [\partial f^{-1}(y)/\partial y]dy$. The choice depends upon the details of the problem and your preference. Regardless of which substitutions are made, the results of an integral evaluation are the same.

For the cases where one of the coordinate variables is a constant, this value is entered, e.g., $y = 4$. The differential for this coordinate $dy = 0$ since there is no variation of the coordinate y on this path.

Step 3: Integrate. The three scalar integrals can be evaluated by setting the integral limits equal to the initial and final values of the appropriate variables, e.g., when dy is the differential in the integral, use $y_{INITIAL}$ and y_{FINAL}. Note that the direction of the path dictates the order of the limits. If the path is reversed, the order of the limits is changed and the integral has an opposite sign. It may seem that a closed path has no initial or final points, but it can be decomposed into two or more separate integrals each of which has an initial and final point based upon the direction of the path.

The first two steps are applicable to both numeric and analytic evaluation of line integrals. However, the procedures for implementing step 3 differ; the next two sections consider the details.

1.11 NUMERIC APPROXIMATION

We have already seen in Eq. (1.22) that an integral along a curved path can be approximated by a sequence of incremental, straight-line segments on which the integral is evaluated. The results of the straight-line evaluations are summed to approximate the integral over the curved path. Moreover, the integral on each segment is further approximated by assuming that the field is constant along the segment; the shorter the segments, the more accurate this approximation.

With this strategy, the integral along a line segment is given by multiplying the field component in the direction of the line segment times the length of the line segment. Length of line segment, integral accuracy, and error limits can be determined by a numerical analysis course. But, we will do it by the following heuristic approach. The evaluation of the field can be made at any convenient point on the line segment. For consistency (and ease of constructing a calculation algorithm), the same relative location can be chosen on each segment, e.g., the initial, final, or midpoint.

This process seems simple enough. Of course we are faced with a tradeoff—the more segments, the more accurately the summation approximates the integral, but the more calculations which must be performed. This common engineering dilemma requires your reasoned and knowledgeable decision in choosing the number of segments. There is no "right answer"; you must choose the "best" solution under the circumstances. What works for one situation may not be the proper choice in another.

Example 1.11-1.   Calculate the line integral of a vector field $\mathbf{F} = x\mathbf{a}_X + y\mathbf{a}_Z$ along the path from (0,0,0) to (1,1,1) via the straight-line segments $(0,0,0) \to (1,0,0) \to (1,1,0) \to (1,1,1)$. First, the dot product is determined as $\mathbf{F} \cdot \mathbf{dl} = x\,dx + y\,dz$; note that $F_Y = 0$, so the dy term vanishes. Since we are approximating the integration by the summation on incremental segments, we replace dx by Δx and dz by Δz. Along the portion of the path from (0,0,0) to (1,0,0), there are no changes in y or z so $\Delta y = \Delta z = 0$; we have only Δx to consider giving $\mathbf{F} \cdot \mathbf{dl} \approx x\Delta x$ on this portion of the path. Along the second portion $\Delta x = \Delta z = 0$ and we have only Δy; but since $F_Y = 0$ there is no contribution. Along the third portion $\Delta x = \Delta y = 0$ and we have only Δz; moreover, along this segment, $y = 1$, so that $\mathbf{F} \cdot \mathbf{dl} \approx \Delta z$. These are combined to approximate the integral along the path as

$$\int_L \mathbf{F} \cdot \mathbf{dl} \approx \underbrace{\sum_{n=1}^{N} x_n \Delta x_n}_{\text{Segment1}} + \underbrace{\sum_{m=1}^{M} \Delta z_m}_{\text{Segment2}}.$$

For computational convenience, let's make the size of all increments within a summation the same, i.e., $\Delta x_n = \Delta x$ and $\Delta z_m = \Delta z$. How large should we choose the increments Δx and Δz? We could get very theoretical about the answer to this question, but on an intuitive basis we know that it depends upon the functional form of the summand. For fields that vary rapidly along the segment, the segments should be smaller; for slowly varying fields, the segments can be larger. The simplest answer is to compare a solution with the path divided into N segments of $\Delta x = L/N$ to the solution with twice as many segments. If the two solutions compare closely, then we probably have a reasonable approximation of the integral.

Since the field of the first summation of Eq. (1.32) is given as x, it varies rather slowly over the range of summation. Let's divide the range into $N = 10$ segments so that $\Delta x = L/N = 1/10 = 0.1$. We must choose the location on each segment at which to evaluate x_n. For convenience, choose the midpoint of each segment, i.e., $x_n = n\Delta x - \Delta x/2 = 0.1n - 0.05$. The summation along this portion of the path is given by

$$\sum_{n=1}^{N} x_n \Delta x_n = \sum_{n=1}^{10} (0.1n - 0.05)(0.1) = (0.01)\left(\sum_{n=1}^{10} n\right) - \sum_{n=1}^{10} 0.005 = 0.5.$$

As a check on the reasonableness of this approximation, we repeat the calculation with 20 segments and obtain $\sum_{n=1}^{N} x_n \Delta x_n = \sum_{n=1}^{20} (0.05n - 0.025)(0.05) = 0.5$ equal to the first result. The exact result (as we will find in the next section) is 0.5. The numeric approximation agrees exactly with the exact solution. Since the limits are in the direction of increasing x, the sign of the integral of segment 1 is positive.

The second summation is seen readily to give $\sum_{m=1}^{M} \Delta z_m = \Delta z_m \sum_{m=1}^{M} 1 = \left(\frac{L}{M}\right) M = L = 1$. The sign of this integral is also positive. Note that this result is exact and independent of the number of segments since the summand is constant. Combining the two summations, we obtain the numeric approximation of the line integral as $\int_{L1} \mathbf{F} \cdot \mathbf{dl} \approx 1.5$ which agrees with the exact result of 1.5.

This technique can also be applied to approximate line integrals where numeric rather than functional values of the field are given as with measured values. The field components can be represented by arrays of numeric values representing the field components at specific locations in space. The appropriate matrix entries are used instead of functional forms to execute the summation procedure.

1.12 ANALYTIC EVALUATION

On many occasions the integrand and the path are expressed as functions of spatial coordinates. Since analytic evaluation of the work integral will give an exact answer, it is often worth the extra effort. The rest of this section includes several examples of analytic solution methods of line integrals.

Example 1.12-1. Calculate the line integral of the vector field given in Example 1.11-1 where $\mathbf{F} = x\mathbf{a}_X + y\mathbf{a}_Z$ along the path from (0,0,0) to (1,1,1) via the straight-line segments (0,0,0,)$\rightarrow$(1,0,0) $\rightarrow$ (1,1,0) $\rightarrow$ (1,1,1). As we found earlier, the dot product is given as $\mathbf{F} \cdot \mathbf{dl} = x\,dx + y\,dz$. Along the first segment, we have only dx and the integrand becomes $x\,dx$. There is no contribution along the second segment. Along the third segment, we have only dz so the integrand becomes dz. The initial and final points are obvious from the specification of the line segments. These are combined to give the integral along the path as

$$\int_{L1} \mathbf{F} \cdot \mathbf{dl} = \int_{x'=0}^{1} x'dx' + \int_{z'=0}^{1} dz' = 1.5.$$

Example 1.12-2. Calculate the line integral for the same field, but for a slightly different path, $(0,0,0) \rightarrow (0,0,1) \rightarrow (0,1,1) \rightarrow (1,1,1)$. The dot product is given by $\mathbf{F} \cdot \mathbf{dl} = x\,dx + y\,dz$ as before. On the first segment $dx = dy = 0$ leaving only $y\,dz$; but since $y = 0$ on this path there is no contribution on this segment. On the second segment $dx = dz = 0$ leaving dy

as the differential; but, its coefficient is zero so this segment gives no contribution. On the third segment $dy = dz = 0$ leaving dx as the differential. These results are combined to give $\int_{L2} \mathbf{F} \cdot \mathbf{dl} = \int_{x'=0}^{1} x' dx' = 0.5$. This result differs from Example 1.12-1; the work integral of this field between the two endpoints is not the same when different paths are used. This means that the field is not conservative and could not represent an electric field intensity.

Example 1.12-3.   Calculate the line integral of the same field of the Examples 1.12-1 and 1.12-2 along a curved path from $(0,0,0)$ to $(1,1,1)$ on the path formed by the intersection of the plane $x = z$ and the parabolic surface $y = \left[\left(x^2 + z^2\right)/2\right]^{1/2}$. As before, the dot product is given by $\mathbf{F} \cdot \mathbf{dl} = x dx + y dz$. The first term requires no further simplification as the functional dependence and the differential are both in terms of a single variable x. In the second term we substitute $x = z$ and $y = \left[\left(x^2 + z^2\right)/2\right]^{1/2}$ which describe the path. This leads to $y dz = \left[\left(x^2 + z^2\right)/2\right]^{1/2} dz = \left[\left(z^2 + z^2\right)/2\right]^{1/2} dz = \pm z dz$. The square root allows two possible solutions, but only $+z dz$ represents the positive values of y that lie on the path. The integral can then be evaluated as $\int_{L3} \mathbf{F} \cdot \mathbf{dl} = \int_{x'=0}^{1} x' dx' + \int_{z'=0}^{1} z' dz' = 1.0$. This result confirms that the field is nonconservative as we have three different results for three different paths.

The electric field intensity within a resistor exhibits conservative behavior; it satisfies KVL and the work integral between any two points is the same regardless of the path chosen. Because the voltage drop is the same for any path, we should seek the path that minimizes the effort in evaluating the integral.

Example 1.12-4.   Find the voltage drop from $(0,0,0)$ to $(1,1,1)$ for an electric field intensity $\mathbf{E} = y\mathbf{a}_X + x\mathbf{a}_Y$. The definition of voltage drop establishes the initial point as $(1,1,1)$ and the final point as $(0,0,0)$. Since an electric field is conservative, we can choose any convenient path to evaluate the integral. Let's choose the segmented path $(1,1,1) \rightarrow (1,1,0) \rightarrow (1,0,0) \rightarrow (0,0,0)$. The dot product is given by $\mathbf{E} \cdot \mathbf{dl} = y dx + x dy$. Since $E_Z = 0$, there is no contribution of the first segment. On the second segment, only $dy \neq 0$ and $x = 1$ so the integrand becomes dy. On the third segment, only $dx \neq 0$; however, since $y = 0$, there is no contribution. The voltage drop is given by

$$V_{0,0,0} - V_{1,1,1} = - \int_{L} \mathbf{E} \cdot \mathbf{dl} = - \int_{y'=1}^{0} dy' = 1\,\text{V}.$$

Example 1.12-5. To verify that **E** of Example 1.12-4 represents a legitimate electric field, let's show that the voltage drop is the same via a different path, say a straight line from the initial point (1,1,1) to the final point (0,0,0). This path is represented by $y = x$ and $y = z$ (or $x = z$). The integrand is $\mathbf{E} \cdot \mathbf{dl} = y\,dx + x\,dy$ as before and the voltage drop is $V_{0,0,0} - V_{1,1,1} = -\int_{x'=1}^{0} x'\,dx' - \int_{y'=1}^{0} y'\,dy' = 0.5 + 0.5 = 1$ V, the same result as in Example 1.12-4.

The voltage drop is the same and KVL is verified for these two paths. This gives us some confidence that **E** does represent a proper electric field intensity. However, these are just two of an infinite number of possible paths for which the voltage drop must be the same. Obviously, we cannot try every possible path; some other verification is necessary to prove that **E** is conservative. You will be pleased to know that there is a single vector operation that unambiguously shows whether a field is conservative. Unfortunately, it requires a new vector operation so we will defer this until later. Nevertheless, in this section, we have learned how to evaluate the work integral over any prescribed path.

Example 1.12-6. Even when electric fields are expressed in non-Cartesian coordinates, the same techniques can be used to evaluate line integrals. Consider the field between two, PEC, coaxial cylinders as shown in Fig. 1.8.

This geometry closely models the common coaxial transmission lines used for cable TV distribution or to interconnect laboratory instruments. Experiments have shown that, with the outer conductor grounded and a voltage V_o applied to the inner conductor, the electric field in the region between the two conductors is given by

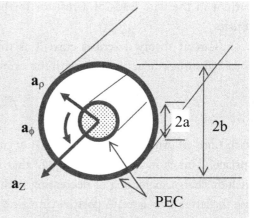

FIGURE 1.8: A coaxial cable.

$\mathbf{E} = \frac{V_o \mathbf{a}_\rho}{\ln(b/a)\rho}$. Note that the electric field points radially outward from the more positive inner conductor to the less positive outer conductor as expected. The differential displacement vector in cylindrical coordinates (see Appendix C) is expressed by $\mathbf{dl} = \mathbf{a}_\rho\,d\rho + \mathbf{a}_\phi\,\rho\,d\phi + \mathbf{a}_z\,dz$; the dot product becomes $\mathbf{E} \cdot \mathbf{dl} = V_o\,d\rho / [\ln(b/a)\,\rho]$. Only displacements in the ρ-direction cause any change in potential since that is the only direction in which **E** points. Displacements in the ϕ- and z-directions are along equipotentials. Since electric fields are conservative, any convenient path can be used to determine the voltage at points between the two conductors.

A radial path is easiest and the voltage is expressed as

$$V(\rho, \phi, z) - V(b, \phi, x) = -\int_{\rho'=b}^{\rho} \frac{V_o d\rho'}{\ln\left(\frac{b}{a}\right)\rho'}$$

$$= -V_o \frac{\ln\left(\frac{\rho}{b}\right)}{\ln\left(\frac{b}{a}\right)} = V_o \frac{\ln\left(\frac{b}{\rho}\right)}{\ln\left(\frac{b}{a}\right)}.$$

For $\rho = a$, the voltage becomes V_o as it must be since this was specified as the applied voltage; for $\rho = b$, the voltage is zero as it must be.

1.13   CURRENT AND CURRENT DENSITY

To this point, we have investigated the electromagnetic concepts associated with the voltage drop across a lumped element resistor. Now it is time to consider another important aspect of resistors—current flow. As with our consideration of voltage, we will relate the measurable current in the wire leads of a resistor to the microscopic details of current flow within the resistor.

Circuit theory describes current as the time rate of flow of charged particles across a specified surface in accordance with the expression

$$I = \dot{Q} \text{ [A]} \qquad (1.33)$$

with units of Amperes. $\dot{Q}$ is the notation to indicate the rate of charge crossing a specified surface. This is in contrast to dQ/dt that represents the rate of change of charge and can include charge increasing or decreasing with a region as well as crossing a boundary. We will use the latter form later. A positive direction of current flow across the surface must be defined also. We can measure the current with an ammeter; the current enters one terminal and leaves the other. Any surface within the ammeter could be considered as the surface through which the current is measured. For convenience, define the surface coincident with an equipotential surface.

According to the passive sign convention of circuits, current flows from the more positive terminal of a resistor to the less positive terminal. The circuit model for current flow describes the flow of positive charges, though, in fact within wires and resistors the vast preponderance of charges in motion are electrons which posses a negative charge . According to Eq. (1.33), electrons (for which $Q < 0$) must flow in the opposite direction of "conventional current" used in circuit theory. The details of the polarity of charge carriers will not trouble us in this course.

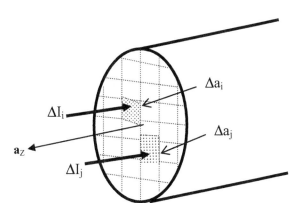

FIGURE 1.9: Current flow in a resistor.

Most often, we will be content to use positive charge flow associated with conventional current of circuit theory with a few rare exceptions in which several different charged particles must be considered.

Current enters the conductive portion of a resistor via the wire leads and, subsequently, the electrodes. As before, the wire leads and electrodes are modeled as having negligibly small resistance. Moreover, as you learned in physics, frequently the current is uniformly distributed across any equipotential surface of an axially symmetric resistor as shown in Fig. 1.9. Using Eqs. (1.1) and (1.2), we can express the voltage drop across a resistor as

$$V_R = \frac{L}{\sigma}\left(\frac{I_R}{A}\right). \tag{1.34}$$

The term in parenthesis is known as *current density* and has units of Amperes/square meter, the smaller the cross-sectional area, the greater the current density and vice versa. Extension of this concept to the incremental current, ΔI_i, crossing the ith surface element of area Δa_i (see Fig. 1.9) leads to current density, J_i, of the ith surface element as a microscopic description of current flow in Amperes/square meter

$$J_i = \frac{\Delta I_i}{\Delta a_i}\ [\text{A/m}^2]. \tag{1.35}$$

For uniform current density, $|\mathbf{J}|$, is simply the total current, I, divided by the resistor cross-sectional area, A. But, in general, the current density can vary throughout the resistor and must be expressed as the quotient of incremental current and incremental area as in Eq. (1.35). As the cross-sectional area is made vanishingly small, the magnitude of the current density at a single point can be expressed as

$$|\mathbf{J}(x, y, z)| = \lim_{\Delta a \to 0} \frac{\Delta I}{\Delta a}. \tag{1.36}$$

Current density has a direction as well since positive charge flows from higher voltage to lower voltage, in the direction of the voltage drop and the electric field intensity. For the resistor of Fig. 1.9, this is in the $-\mathbf{a}_Z$ direction. Note that $\pm\mathbf{a}_Z$ is normal to the incremental surface used to define current density, Δa. It is standard practice to consider the current density as a vector and to attach to it the unit vector in the direction of positive charge flow as

$$\mathbf{J} = \frac{\Delta I_Z}{\Delta a}\mathbf{a}_Z. \qquad (1.37)$$

The current density parameter, $\mathbf{J}$, contains more information—microscopic information—regarding current flow than provided by total current. Earlier we obtained $\mathbf{E}$, an intensity vector of volts/meter. Now we have defined a different form of vector called a flux density vector with units of Amperes/square meter or Coulombs/second/square meter, $\mathbf{J}$, in the same direction as $\mathbf{E}$ for our resistor. Flux density represents the directed strength or magnitude of a vector that penetrates or crosses a unit surface perpendicular to the vector and has units of flux/square meter. The flow of water within a region could be represented by a flux density vector $\mathbf{W}$ given in liters/second/square meter and pointing in the direction of the flow. For a nuclear reactor, the flux of neutrons/second/square meter could be represented by $\dot{\mathbf{N}}$. All vector fields of electromagnetics are either an intensity vector or a flux density vector. While $\mathbf{J}$, $\mathbf{W}$, and $\dot{\mathbf{N}}$ represent the actual motion of physical objects or mass, some flux densities represent invisible, vector fields, fixed in space, which are not associated with motion of anything physical. We will discuss this more in future chapters.

Example 1.13-1.   A #16 copper wire has a current of 10 A flowing in the axial direction. Calculate the current density within the wire. Let's assume that the current is uniformly distributed. Since the current flow is axial, it is perpendicular to the diameter. Therefore, current density is simply the current divided by the area. The diameter of the #16 wire is 0.0508″ with an area of $A = \pi[(0.0254)(0.0254)]^2 = 1.31 \times 10^{-6}$ m^2 so that $|\mathbf{J}| = 10/1.31 \times 10^{-6} = 7.65$ MA/m^2. Wow! What a large value, but it is typical of the current density in house wiring.

1.14   CURRENTS AND SURFACE INTEGRALS

For most common materials, $\mathbf{J}$ is in the direction of $\mathbf{E}$ which in turn is perpendicular to equipotential surfaces. The total current crossing an equipotential surface in a resistor is found

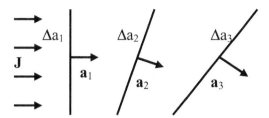

FIGURE 1.10: Current flow through surfaces.

by summing all the incremental currents which perpendicularly cross this surface as

$$I = \sum_{i=1}^{M} \Delta I_i = \sum_{i=1}^{M} J_i \Delta a_i. \tag{1.38}$$

In more general cases we may wish to calculate the current crossing surfaces other than equipotentials to which the flow is not perpendicular. How can Eq. (1.38) be adapted to handle this?

Consider the surfaces of Fig. 1.10; the same total current passes through each of them even though they have quite different areas. A little thought suggests that, in calculating the total current through a surface, the really important surface parameter is projected area—the area projected onto a plane perpendicular to the current flow. Fortunately, this is relatively easy to find in terms of the unit normal vector to the surface. Every surface has two vectors which are perpendicular, or normal, to it. We must choose the normal vector in the direction that we wish to define for positive current flow. A directed surface element is then defined as

$$\Delta \mathbf{s} = \mathbf{a}_N \Delta a \tag{1.39}$$

in which the incremental area multiplies the unit normal vector. Remember, that when we specify the equation of a surface, its normal can be expressed as its gradient, see Eq. (1.19). From Eq. (1.39), the projected area of the incremental surface element perpendicular to the current flow is given as

$$\Delta a_J = \mathbf{a}_J \cdot \Delta \mathbf{s} = \mathbf{a}_J \cdot \mathbf{a}_N \Delta a \tag{1.40}$$

As in Eq. (1.38), the current crossing the incremental surface is found by multiplying the projected area by the magnitude of the current density

$$\Delta I = |\mathbf{J}| \Delta a_J = \mathbf{J} \cdot \Delta \mathbf{s} = |\mathbf{J}| \mathbf{a}_J \cdot \mathbf{a}_N \Delta a$$
$$= J_N \Delta a. \tag{1.41}$$

How elegant! Only the component of current density that is perpendicular to the surface passes through the surface; the parallel component does not cross the surface. The incremental current through an incremental surface element is merely the dot product of the current density, **J**, and the directed surface element, Δ**s**. More generally, we can replace the current flux density, **J**, by an arbitrary flux density, **F**. This results in an incremental flux, $\Delta\Psi$, that corresponds to the incremental current, ΔI, crossing the surface, Δ**s**. This is expressed as

$$\Delta\Psi = \mathbf{F} \cdot \Delta\mathbf{s}. \qquad (1.42)$$

These concepts enable a modification to the form of Eq. (1.38) so that it is valid for any surface

$$I = \sum_{i=1}^{M} \Delta I_i = \sum_{i=1}^{M} \mathbf{J}_i \cdot \Delta\mathbf{s}_i \qquad (1.43)$$

Recall that when summing the incremental work done along a particular path, we took the limit as the increments became differentially small which enabled us to represent the work done as a line integral. Similarly, if we let the incremental surface of Eq. (1.43) become differentially small, we can obtain an integral representation for the current crossing a surface

$$I = \iint_S \mathbf{J} \cdot \mathbf{ds} \qquad (1.44)$$

where S is the extent of the finite surface. The double integral is the usual notation for the integral over a surface. This form of integral is known as a *flux integral* since it describes the flux, I, crossing the surface, S. In generalized form this flux integral becomes

$$\Psi = \iint_S \mathbf{F} \cdot \mathbf{ds}. \qquad (1.45)$$

This is the second type of vector integral that we will use throughout our study of electromagnetics. Master it! The physical meaning contained within Eq. (1.44) is the subject of the next section.

1.15   CHARGE CONSERVATION AND KCL

Intuitively, we recognize that charge is conserved within any volume. If current flows in, the positive charge must increase; if current flows out, it must decrease. To express this relationship mathematically, we can use a form of Eq. (1.44). The current entering or leaving a volume must cross the surface that encloses the volume. This is called a closed surface since it has "no holes in it," otherwise the volume would not be enclosed by the surface. The net current out of the

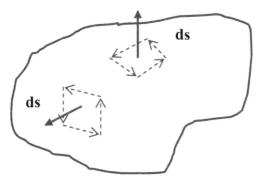

FIGURE 1.11: Geometry of a closed surface.

closed surface S is represented mathematically by

$$I_{\text{OUT}} = \oiint_S \mathbf{J} \cdot \mathbf{ds} \qquad (1.46)$$

where S is the surface which encloses the volume and **ds** points *out* of the volume, see Fig. 1.11. The direction of **ds** is in the direction of the thumb when the *fingers of the right hand* describe a path that encloses an incremental area on the surface. The arrows show the direction of the fingers. The choice of the outward normal to the surface means that components of **J** which point outward will contribute positively to the current, those which point inward will contribute negatively. To calculate the net current into a volume, the inward directed **ds** must be chosen. The closed loop on the integral symbol is the mathematician's way of showing a closed surface much as it indicated a closed line integral. Recall that current represents the rate of charge crossing a particular surface. In this case, it is the surface enclosing a particular volume and is equal to the rate of charge crossing the surface. If charge within the volume is conserved, the rate of charge moving outward across the closed surface S must be equal to the rate at which the charge within the volume is decreasing. With the definition of the net charge within the volume as Q, charge conservation can be expressed as

$$I_{\text{OUT}} = \oiint_S \mathbf{J} \cdot \mathbf{ds} = -\frac{dQ}{dt}. \qquad (1.47)$$

The greater the outward components of the current density, **J**, the greater the rate of decrease of charge within the volume. Note that Q is the net charge within the volume enclosed by the closed surface S, i.e., the algebraic sum of all charges, both positive and negative.

We are already familiar with this concept in circuit theory for the special case where there is no net charge within the volume. It is known as Kirchoff's current law (KCL). If there is no net charge within the volume, then Q is zero and unchanging with a zero derivative, i.e.,

$dQ/dt = 0$. Under these conditions, Eq. (1.47) is expressed as

$$I_{OUT} = \oiint_S \mathbf{J} \cdot \mathbf{ds} = 0 \tag{1.48}$$

which we interpret to say that the net current out of any surface sums to zero. In circuits we usually focused upon a single node, but we could combine any number of nodes to define the surface S. However, we will later see that the surface S must completely contain an element, not just a portion of it, in order that Eq. (1.48) is valid. So you already know more about surface integrals than you realize! But, on to something new, we must learn how to evaluate surface integrals.

1.16 EVALUATION OF SURFACE INTEGRALS

Our earlier experience with evaluation of line integrals provides several useful parallels for evaluation of surface integrals. Both integrands are formed as a dot product and both integrations are confined to a specified region. Of course, they have quite different physical meanings. Line integrals involve a field intensity vector integrated in a specified direction along the given path; they are proportional to work done or to change in energy. Surface integrals involve a flux density vector integrated over a specified region of a given directed surface; they are equal to the total flux that passes through the surface in a specified direction.

As with line integrals, just three steps are involved in the evaluation of a surface integral. First, perform the dot product of the integrand and the differential to form the scalar differential flux; second, evaluate this scalar for the surface over which the integration is to take place; third, perform the scalar integration over the limits of the defined surface S.

The directed surface element $\mathbf{ds} = \mathbf{a}_N da$ is the product of two factors—the surface normal vector and the associated differential area. The proper surface normal is defined to be in the direction in which the desired flux is to be calculated. If the direction of the normal is reversed, the sign of the integral is changed. The convention for closed surfaces is that the outward surface normal is chosen so that the integral represents the flux out of the closed surface. The associated differential area is the product of two differential lengths which lie on the surface S. A careful look at Fig. 1.12 reveals a compact representation of the directed surface normal as

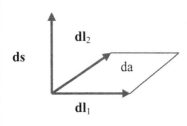

FIGURE 1.12: Directed differential surface element.

$$\mathbf{ds} = \mathbf{dl}_1 \times \mathbf{dl}_2. \tag{1.49}$$

As you recall, the cross product of two vectors is perpendicular to the plane containing the two vectors and is proportional to the area of the parallelogram formed by the two vectors, see Appendix B. dl_1 and dl_2 must be chosen carefully so that ds is oriented in the desired direction. Usually, we need not be so formal as Eq. (1.49) since the surface normal and the differential area are obvious from the surface geometry of S.

The choice of coordinate system in which to express the directed surface element is guided most often by convenience. When the surface (or a portion of it) conforms to a common coordinate system, e.g., Cartesian, cylindrical, or spherical, the choice of coordinate system is trivial. However, rarely does the desired surface conveniently fit a common coordinate system. Several techniques for handling these cases are included in the examples that follow.

With both J and ds expressed in terms of the same coordinate system, the dot product of $J \cdot ds$ is readily formed. Integration can proceed only after $J \cdot ds$ is evaluated on the surface S. As noted earlier, a single equation is required to represent a surface. For those cases where a common coordinate system is appropriate, the equation often takes the form of a single variable, e.g., $x = 5$ represents a planar surface whereas a spherical surface is given as $r = 4$. The general representation of a surface is as a single function expressed in terms of appropriate coordinates. For the common coordinate systems, expressions for the surface can be written in the form of $S(x, y, z) = C_1$, $S(\rho, \phi, z) = C_2$, or $S(r, \theta, \phi) = C_3$, where the C_i's represent constants. In each representation, any one variable can be expressed in terms of the other two when on the surface—an expression unique for that surface. Substitution of this expression into the dot product, $J \cdot ds$, provides the required evaluation on the surface.

In contrast to line integrals where the initial and final points of integration are quite clear, surface integrals require care in setting the limits of integration. This is easily handled by considering a "test" flux density of unit magnitude that is perpendicular to the surface. The dot product becomes $J \cdot ds = 1a_N \cdot a_N da = da$; the resulting surface integral is the area of the surface S. Since surface areas are always positive, the limits of integration are set properly when the "test" integral gives a positive number. This usually means that the upper limit of integration of each variable is more positive than the lower limit.

1.17 NUMERIC EVALUATION

Surface integrals can be evaluated numerically in a manner similar to that used with line integrals. The simplest approach is to discretize the surface S into a set of incremental areas each of which is then approximated by the tangent plane at the center of the region. Furthermore, if the incremental areas are small enough, variations of the integrand within each area are ignored and the integrand is assumed to be constant and equal to the value at the center of the region. As the size of the areas is decreased, the accuracy of these approximations becomes greater, but the number of regions and calculations increases. We are faced with a typical engineering decision—accuracy versus computational effort. Since we are primarily interested in the nature of the behavior of electromagnetic fields, not accurate to many decimal places, we will be satisfied with this simple approximation. More sophisticated numeric techniques are needed for increased accuracy and computational efficiency. They are the subject of courses in numeric analysis and advanced electromagnetics courses.

This approach allows us to approximate the surface integral of Eq. (1.44) by Eq. (1.43) as

$$I = \iint_S \mathbf{J} \cdot \mathbf{ds} \approx \sum_{i=1}^{M} \mathbf{J}_i \cdot \Delta\mathbf{s}_i. \qquad (1.50)$$

$\mathbf{J}_i$ represents the numeric value of the vector flux density at the center of the ith incremental tangent plane, $\Delta\mathbf{s}_i$. A form of Eq. (1.50) more useful for calculations is

$$I = \sum_{j=1}^{M}\sum_{i=1}^{N} \mathbf{J}_{ij} \cdot \mathbf{a}_{Nij}\Delta l_{1i}\Delta l_{2j} \qquad (1.51)$$

where i and j represent the summation indices of the two incremental lengths, Δl_{1i} and Δl_{2j}, $\mathbf{J}_{ij}$ is the value of the current flux density at the center of the surface element denoted by the ith value of Δl_1 and the jth value of Δl_2 and $\mathbf{a}_{Nij}$ is the surface normal at the center of the ij surface element. For flux densities expressed in analytic form, $\mathbf{J}_{ij}$ is merely the equation for $\mathbf{J}$ evaluated at the center of the region. Alternatively, $\mathbf{J}_{ij}$ may be a table of numeric values representing sampled or measured values of flux density. To evaluate $\Delta\mathbf{s}_i$, first find the surface normal at the center of the region. Since S represents a surface and is expressed as a function of coordinates, the surface normal can be expressed in terms of the gradient, described in Eq. (1.19), as

$$\mathbf{a}_{Nij} = \frac{\nabla S_{ij}}{|\nabla S_{ij}|} \qquad (1.52)$$

which is evaluated at the center of the ij surface element. Unless the surface S naturally fits some other coordinate system, $\mathbf{a}_{Nij}$ is most frequently expressed in terms of the Cartesian coordinate system with invariant unit vectors throughout space. The incremental area $\Delta a_{ij} = \Delta l_{1i}\Delta l_{2j}$ is expressed in a convenient coordinate system, most often Cartesian. Regardless of the coordinate system, Δa_{ij} is evaluated at the center of the region also. Following these steps, the dot product and subsequent summation are taken. As with line integrals, it is convenient to choose equal increments of the surface variables, i.e., $\Delta l_{1i} = \Delta l_1$ and $\Delta l_{2j} = \Delta l_2$. As with line integrals, the number of intervals affects the accuracy in a way which depends upon the functional variations of the incremental flux over the surface S. The more rapid the variations, the more intervals needed. If the number of intervals is doubled and the results do not change markedly, then the number of intervals is probably sufficient.

The several examples which follow will illustrate the concepts discussed in this section.

Example 1.17-1. Calculate the total current flux in the positive y-direction which passes through the region of the $y = 0$ plane defined by $0 \leq x \leq 1$ and $0 \leq z \leq 1$ for a flux density of $\mathbf{J} = 5y\mathbf{a}_X - 3y\mathbf{a}_Y$. The desired direction of flux is the positive y-direction, so $\mathbf{a}_{Nij} = \mathbf{a}_Y$; the differential area lies in the $y = 0$ plane so that $\Delta a = \Delta x \Delta z$ and $\Delta \mathbf{s} = \mathbf{a}_Y \Delta x \Delta z$. The dot product is $\mathbf{J} \cdot \Delta \mathbf{s} = (5y\mathbf{a}_X - 3y\mathbf{a}_Y) \cdot \mathbf{a}_Y \Delta x \Delta z = -3y\Delta x \Delta z$, which vanishes on the surface S where $y = 0$. Since the incremental flux vanishes, no flux passes through the surface.

Example 1.17-2. Consider Example 1.17-1, but choose the surface S as the $y = 2$ plane. As before, we find $\mathbf{J} \cdot \Delta \mathbf{s} = -3y\Delta x \Delta z$ which is nonzero on the $y = 2$ plane. Consequently, the incremental flux is a constant $\Delta I = \mathbf{J} \cdot \Delta \mathbf{s}|_{y=2} = -6\Delta x \Delta z$ over the entire surface. With no variation of ΔI over the surface, a single increment sufficient, i.e., $\Delta x = \Delta z = 1$, with the result $I = -6 \sum_{i=1}^{1} \sum_{j=1}^{1} (1)(1) = -6$ A.

Example 1.17-3. Consider the current flux density of Example 1.17-1; calculate the total current in the positive x-direction crossing the $x = 2$ surface within the range of $0 \leq y \leq 1$ and $0 \leq z \leq 1$. The surface normal is $\mathbf{a}_N = \mathbf{a}_X$; the incremental area is $\Delta a = \Delta y \Delta z$ so that $\Delta I = \mathbf{J} \cdot \Delta \mathbf{s} = (5y\mathbf{a}_X - 3y\mathbf{a}_Y) \cdot \mathbf{a}_X \Delta y \Delta z = 5y\Delta y \Delta z$. Since ΔI varies with respect to y but not with z; we choose the number of increments as $N_Y = 10$ and $N_z = 1$, which gives $y_j = 0.1(j - 0.5)$, $\Delta y = 0.1$, and $\Delta z = 1$. This leads to

$$I = \sum_{j-1}^{10} \sum_{i=1}^{1} 5(0.1)(j - 0.5)(0.1)(1) = 0.05 \sum_{j=1}^{10} (j - 0.5)$$

$$= 0.05(55 - 5) = 2.5 \text{ A}.$$

Doubling N_Y leads to $y = 0.05(j - 0.5)$ and $\Delta y = 0.05$ but since there are no z variations, N_Z is unchanged and $\Delta z = 1$ with the result $I = \sum_{j=1}^{20} \sum_{i=1}^{1} 5(0.05)(j - 0.5)(0.05)(1) = 2.5$ which agrees exactly with the result for half the number of increments. This gives us confidence that we have sufficiently small increments. As we will see later, both agree with the analytic calculations due to the linear variation of the incremental current. We should not expect such a good agreement with other functional forms.

Example 1.17-4. A point source of light located at the origin emanates power equally in all directions, i.e., it is isotropic. At a radius of 1 m, this light flux density is 1 W/m². Calculate the light power that emanates radially outward through the upper hemispherical surface. For convenience, we will use the spherical coordinate system with the sphere centered on the coordinate origin. From the description of light power density we can determine that $\mathbf{F} = 1\mathbf{a}_r$ on the surface where $\mathbf{a}_r$ is the radial unit vector of spherical coordinates. The direction in which light power flow is desired is the radially outward direction through the spherical surface so that $\mathbf{a}_N = \mathbf{a}_r$. The surface area can be expressed in terms of incremental lengths in the θ and ϕ directions. As you recall (or have found in Appendix B), the lengths can be expressed as $\Delta l_1 = r\Delta\theta$ and $\Delta l_2 = r\sin\theta\Delta\phi$, where $r = 1$ on the unit sphere surface. This functional dependence of the incremental length will complicate the calculations. Combining these results, we obtain the incremental power flux density $\Delta\Psi = \mathbf{F} \cdot \Delta\mathbf{s} = \mathbf{a}_r \cdot \mathbf{a}_r(r\Delta\theta)(r\sin\theta\Delta\phi)|_{r=1} = \sin\theta\Delta\theta\Delta\phi$. To cover the upper hemisphere, the range is $0 \le \theta \le \pi/2$ and $0 \le \phi \le 2\pi$ (refer to Appendix B). $\Delta\Psi$ does not vary with ϕ so that $N_\phi = 1$ is sufficient; but since $\Delta\Psi$ varies as $\sin\theta$, choose $N_\theta = 10$ so that $\Delta\theta = \pi/20$ and $\theta_j = (j - 0.5)\Delta\theta = (j - 0.5)\pi/20$. The total flux is expressed as

$$\Psi = \sum_{i=1}^{1}\sum_{j=1}^{10} \sin[\theta_j]\Delta\theta_j\Delta\phi_i$$

$$= \sum_{i=1}^{1}\sum_{j=1}^{10} \sin\left[(j - 0.5)\left(\frac{\pi}{20}\right)\right]\left(\frac{\pi}{20}\right)(2\pi)$$

$$= \left(\frac{\pi^2}{10}\right)\sum_{j=1}^{10} \sin\left[(j - 0.5)\left(\frac{\pi}{20}\right)\right] = 6.2896 \text{ W}.$$

As a check on the accuracy of our result set $N_\theta = 20$ which gives $\Psi = \sum_{i=1}^{1}\sum_{j=1}^{20} \sin[(j - 0.5)(\frac{\pi}{40})](\frac{\pi}{40})(2\pi) = 6.2848$ W, close enough to the previous value to consider the solution accurate. In fact, the analytic solution is 2π so these simple approximations are quite accurate. This calculation means that 2π Watts of light power is emanating through the upper hemisphere. Since the source is isotropic, an equal amount of power is emanating from the bottom hemisphere; the total power from the sphere is 4π Watts.

Example 1.17-5. An isotropic current flux density of 1 A/m² is directed radially across the spherical surface of unit radius. Describe the conditions of the charge Q contained within this surface. The total current is found to be 4π Amperes using the identical calculations as the power of Example 1.17-4. Conservation of charge requires that when there is a net current flow out of a closed surface, the charge within the surface must be decreasing. Equation (1.47) expresses this quantitatively as $\frac{dQ}{dt} = -I_{OUT} = -4\pi$ C/s. The charge within the spherical surface is decreasing at a constant rate of 4π Coulombs per second due to the current out of the sphere of 4π Amperes. The charge within the volume is found by direct integration to be $Q(t) = Q(0) - 4\pi t$. The charge present in the volume at $t = 0$, $Q(0)$ must be specified to get a complete solution.

Example 1.17-6.   The same isotropic current density of Example 1.17-5 passes through the surface of cube with sides of length 2 m centered on the origin oriented with its faces parallel to the x, y, and z planes. An additional detail regarding the flux is that it is inversely proportional to the square of the distance from the point source. Note that since the sphere is everywhere equidistant from the origin, the flux density is uniform so this feature was unnecessary when the source of current was centered within the sphere. $\mathbf{J}$ is expressed naturally in spherical coordinates as $\mathbf{J} = r^{-2}\mathbf{a}_r$ A/m^2. The sides of the box have outward surface normals of $\pm\mathbf{a}_X$, $\pm\mathbf{a}_Y$, and $\pm\mathbf{a}_Z$ for the front, back, right, left, top, and bottom sides, respectively. To simplify the dot product of incremental flux, convert $\mathbf{a}_r$ to its rectangular components $\mathbf{a}_r = \sin\theta\cos\phi\mathbf{a}_X + \sin\theta\sin\phi\mathbf{a}_Y + \cos\theta\mathbf{a}_Z$ (with a little help from Appendix B).

First, let's find the flux through the top surface of the cube where $z = 1$, $\mathbf{a}_N = \mathbf{a}_Z$, $\Delta a = \Delta x\Delta y$, and the range of integration is $-1 \leq x \leq 1$ and $-1 \leq y \leq 1$. The incremental flux is given by $\Delta\psi = \mathbf{J} \cdot \Delta\mathbf{s} = \mathbf{a}_r \cdot \mathbf{a}_Z\Delta x\Delta y/r^2 = \cos\theta\,\Delta x\Delta y/r^2$. We convert the spherical coordinate variables to Cartesian coordinates from their definitions in Appendix B as $1/r^2 = 1/(x^2 + y^2 + z^2)$ and $\cos\theta = z/(x^2 + y^2 + z^2)^{1/2}$. The symmetry of this problem allows a significant reduction in the number of calculations. The flux which penetrates the $z = 1$ surface is the same for all four quadrants of the $z = 1$ surface. Consequently, the flux through the top surface is obtained by integrating over the range of $0 \leq x \leq 1$ and $0 \leq y \leq 1$ and multiplying the result by 4, resulting in a four-fold reduction in the number of calculations. Since the incremental flux varies over the surface, choose $N_X = N_Y = 10$ so that $\Delta x = \Delta y = 0.1$. Combining these results, we find the current flux out of the $z = 1$ surface as

$$I = \sum_{i=1}^{10}\sum_{j=1}^{10} \frac{4(0.1)(0.1)}{(1 + (0.1)^2[(i - 0.5)^2 + (j - 0.5)^2])^{3/2}}$$

$$= 4(0.01)(52.392) = 2.09568 \; A.$$

A similar calculation for the bottom surface at $z = -1$ gives the same result due to the isotropic nature of $\mathbf{J}$. In fact, a little thought suggests that due the symmetry of $\mathbf{J}$, the fluxes through each of the surfaces of the cube are equal. To verify this reasoning, let's calculate the flux through the right surface at $x = 1$, where $\mathbf{a}_N = \mathbf{a}_X$, $\Delta a = \Delta y \Delta z$, and $\Delta I = \mathbf{J} \cdot \Delta \mathbf{s} = \sin\theta \cos\phi / r^2$ over the range of integration of $0 \leq y \leq 1$ and $0 \leq z \leq 1$. From the relationships, $\sin\theta = [(x^2 + y^2)/(x^2 + y^2 + z^2)]^{1/2}$ and $\cos\phi = x/(x^2 + y^2)^{1/2}$, then $\sin\theta \cos\phi = x/(x^2 + y^2 + z^2)^{1/2}$, the same form as for the $z = 1$ surface. Of course, the numeric integration gives the same value we got above. Now, we can express confidently the total current out of the cube is just six times the current through the top surface, or $I = 6 I_{TOP} = 6(2.0957) = 12.574 \approx 4\pi$ A. Not so amazingly, this is the same result that we obtained for the flux emanating from a unit sphere for the same flux density. This verifies that the same total current emanates from any closed surface that includes the same source regardless of its shape.

1.18   ANALYTIC EVALUATION

When the flux density and the surface are expressed in functional form, standard analytic integration techniques can be used to calculate the flux surface integral. This method is most useful when the surface conforms to one of the common coordinate systems. Less frequently, a common surface over which an integral can be readily evaluated, e.g., a sphere, a cylinder, or a cone, is used to approximate a surface which cannot be integrated easily. The same "three-step" procedure used with line integrals is followed here as well. Several examples that follow will illustrate the method.

Example 1.18-1.   Reconsider Example 1.17-4 with isotropic current flowing through a spherical surface. The current density is given as $\mathbf{J} = (1/r)^2 \mathbf{a}_r$ A/m^2; the surface normal is $\mathbf{a}_N = \mathbf{a}_r$; the differential area is $da = r^2 \sin\theta d\theta d\phi$. Combining these, we can express the current flowing out of the sphere as

$$I = \oiint_S \mathbf{J} \cdot \mathbf{ds} = \int_{\phi=0}^{2\pi} \int_{\theta=0}^{\pi} \frac{\mathbf{a}_r}{r^2} \cdot \mathbf{a}_r r^2 \sin\theta d\theta d\phi$$

$$= \left(\int_{\phi=0}^{2\pi} d\phi \right) \left(\int_{\theta=0}^{\pi} \sin\theta d\theta \right) = (2\pi)(2) = 4\pi \ A.$$

As we anticipated, the numeric integration agrees with the analytic integration. More significantly, note that the integral shows no radial dependence. This means that regardless of what radius we choose, the current is the same, i.e., the current is conserved over any surface which encloses the source. The current density's inverse square dependence upon radius ensures this behavior. The radial dependence of many physical fields has this form as well. More on this will be discussed later.

Example 1.18-2. A nonuniform current density, $\mathbf{J} = (\sin\theta/r^2)\mathbf{a}_r$, passes through a unit sphere. This current is zero at the poles of the sphere where $\sin\theta = 0$ and peaked in the equatorial plane where $\sin\theta = \pi/2$. Calculate the total current outflow. The integral is formulated as in Example 1.18-1, but the integrand is complicated by the additional $\sin\theta$ term; the integral is expressed as

$$I = \oiint_S \mathbf{J} \cdot \mathbf{ds} = \left(\int_{\phi=0}^{2\pi} d\phi \right) \left(\int_{\theta=0}^{\pi} \sin^2\theta\, d\theta \right) = (2\pi)\left(\frac{\pi}{2}\right) = \pi^2\ A.$$

Other functional forms for current density are handled in the same manner, though the integrals may not always be as easy to evaluate.

Example 1.18-3. All of space is filled with a flux density expressed in cylindrical coordinates as $\mathbf{F} = \rho^2 \cos\phi\,\mathbf{a}_\rho + 5\rho z \sin\phi\,\mathbf{a}_\phi + 3\rho\mathbf{a}_Z$ w/m^2 where w is the symbol for widgets. Calculate the flux of widgets emanating from the volume shown in Fig. 1.13.

We can write an expression for the total flux as $\Psi = \oiint_S \mathbf{F} \cdot \mathbf{ds}$. We know how to calculate the integral, but we must carefully define the several parts of the surface S. The curved surface is defined as $S_C\{\rho = 2, 0 \leq \phi \leq \pi/4$, and $-1 \leq z \leq 3\}$, the left end as $S_L\{z = -1, 0 \leq \phi \leq \pi/4$, and $0 \leq \rho \leq 2\}$, the right end as $S_R\{z = 3, 0 \leq \phi \leq \pi/4$, and $0 \leq \rho \leq 2\}$, the horizontal surface as $S_H\{\phi = 0, 0 \leq \rho \leq 2$, and $-1 \leq z \leq 3\}$, and the remaining inclined surface as $S_I\{\phi = \pi/4, 0 \leq \rho \leq 2$, and $-1 \leq z \leq 3\}$. This seems quite

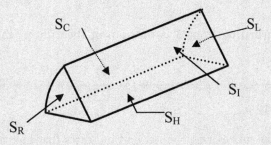

FIGURE 1.13: Surfaces of pie-shaped cylinder.

a bit more complicated than in earlier examples since five surface integrals must be evaluated. But, each of them follows the same three rules that we used before. For each surface, we must identify the directed surface element, form the differential flux dot product, set the limits on the integral, and, finally, integrate. So let's begin.

$$S_C : \mathbf{ds} = \mathbf{a}_\rho dz\rho d\phi,$$

$$d\Psi_C = \mathbf{F} \cdot \mathbf{ds}|_{\rho=2} = (\rho^2 \cos\phi)dz\rho d\phi|_{\rho=2}$$

$$= 8\cos\phi dz d\phi, \text{ and}$$

$$\Psi_C = 8 \int_{z=-1}^{3} dz \int_{\phi=0}^{\pi/4} \cos\phi d\phi = 8(4)\left(\frac{1}{\sqrt{2}}\right) = \frac{32}{\sqrt{2}} w.$$

$$S_L : \mathbf{ds} = -\mathbf{a}_z d\rho\rho d\phi,$$

$$d\Psi_L = \mathbf{F} \cdot \mathbf{ds}|_{z=-1} = -3\rho^2 d\rho d\phi, \text{ and}$$

$$\Psi_L = -3 \int_{\rho=0}^{2} \rho^2 d\rho \int_{\phi=0}^{\pi/4} d\phi = -3\left(\frac{8}{3}\right)\left(\frac{\pi}{4}\right) = -2\pi \ w.$$

$$S_R : \mathbf{ds} = \mathbf{a}_z d\rho\rho d\phi,$$

$$d\Psi_L = \mathbf{F} \cdot \mathbf{ds}|_{z=3} = 3\rho^2 d\rho d\phi, \text{ and}$$

$$\Psi_R = 3 \int_{\rho=0}^{2} \rho^2 d\rho \int_{\phi=0}^{\pi/4} d\phi = 3\left(\frac{8}{3}\right)\left(\frac{\pi}{4}\right) = 2\pi \ w.$$

$$S_H : \mathbf{ds} = -\mathbf{a}_\phi d\rho dz,$$

$$d\Psi_I = \mathbf{F} \cdot \mathbf{ds}|_{\phi=0} = -5\sin\phi\rho d\rho z dz|_{\phi=0} = 0, \text{ and}$$

$$\Psi_H = 0 \ w.$$

$$S_I : \mathbf{ds} = \mathbf{a}_\phi d\rho dz,$$

$$d\Psi_I = \mathbf{F} \cdot \mathbf{ds}|_{\phi=\pi/4} = 5\rho z \sin\phi d\rho dz|_{\phi=\pi/4}$$

$$= (5/\sqrt{2})\rho d\rho z dz, \text{ and}$$

$$\Psi_I = \frac{5}{\sqrt{2}} \int_{\rho=0}^{2} \rho d\rho \int_{z=-1}^{3} z dz = \frac{5}{\sqrt{2}}(2)(4) = \frac{40}{\sqrt{2}} w.$$

The total flux emanating from S is the sum of these fluxes through the individual surfaces

$$\Psi = \Psi_C + \Psi_L + \Psi_R + \Psi_H + \Psi_I = \frac{32}{\sqrt{2}} - 2\pi + 2\pi + 0 + \frac{40}{\sqrt{2}}$$

$$= 50.912 \ w.$$

The negative sign for the flux exiting the left end indicates that flux actually enters the left end. Moreover, the flux entering the left end is equal to the flux exiting the right end. This is due to the fact that the z-component of the flux has no dependence upon z so that the same flux density exists at both ends which are of equal area; hence, equal flux entering one end as leaving the other.

1.19 DIVERGENCE

Surface integrals provide "global" information about the total flux emanating from a finite-sized surface. For many situations this is adequate as in KCL of circuit nodal analysis. However, it would often be useful to know the behavior of the flux at a point. Theoretically, this would be on a differential scale, but more practically, we can only observe or perform numeric calculations on an incremental scale. To accomplish this, we apply the flux integral to the surface S shown in Fig. 1.14, evaluating it as the dimensions of the surface are allowed to shrink to incrementally small lengths. Using the convention that a closed surface integral represents the flux out of the region, we find the current flux out of the closed surface S is given by Eq. (1.46) as

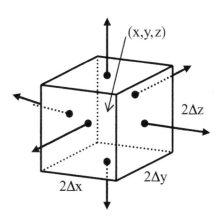

FIGURE 1.14: Flux emanating from an incremental surface S.

$$I = \oiint_S \mathbf{J} \cdot \mathbf{ds} \tag{1.53}$$

For convenience, the volume is assumed to be a rectangular box centered at the point (x, y, z) with the surfaces perpendicular to the coordinate axes. The current density is written in terms of its Cartesian components as

$$\mathbf{J} = J_X \mathbf{a}_X + J_Y \mathbf{a}_Y + J_Z \mathbf{a}_Z \tag{1.54}$$

where the components J_i may be functions of space. The total current out is composed of six currents crossing the front, back, right, left, top, and bottom surfaces of the box. Since the box becomes vanishingly small, the total current is indicated as an incremental current and is expressed as

$$\Delta I = \Delta I_F + \Delta I_{\text{BACK}} + \Delta I_R + \Delta I_L + \Delta I_T + \Delta I_{\text{BOT}}. \tag{1.55}$$

The current out of the front and back surfaces can be combined as

$$\Delta I_F + \Delta I_{\text{BACK}} = \iint_{\text{FRONT}} \mathbf{J} \cdot \mathbf{ds} + \iint_{\text{BACK}} \mathbf{J} \cdot \mathbf{ds}, \tag{1.56}$$

since $\mathbf{ds} = \mathbf{a}_X dy\,dz$ and $\mathbf{ds} = -\mathbf{a}_X dy\,dz$ for front and back surfaces, respectively. These currents are due only to the x-component of the current density, J_X. Moreover, the differential surface elements are the same for both surfaces, $dy\,dz$. In addition, the well-known Taylor's series can represent J_X on both surfaces in terms of J_X at the center of the box as

$$J_X\left(x \pm \frac{\Delta x}{2}, y, z\right) = J_X(x, y, z)$$
$$+ \frac{\partial J_X(x, y, z)}{\partial x}\frac{(\pm \Delta x)}{2!}$$
$$+ \frac{\partial^2 J_X(x, y, z)}{\partial x^2}\frac{(\pm \Delta x)^2}{3!} + \cdots \qquad (1.57)$$

where the "+" sign applies for the front surface and the "−" sign for the back. Substituting Eq. (1.57) into Eq. (1.56), we have

$$\Delta I_F + \Delta I_{BACK} = \underset{FRONT}{\iint} \mathbf{J} \cdot \mathbf{ds} + \underset{BACK}{\iint} \mathbf{J} \cdot \mathbf{ds}$$
$$= \underset{\Delta z\, \Delta y}{\int \int}\underbrace{\left(J_X + \frac{\partial J_X}{\partial x}\frac{\Delta x}{2!} + \frac{\partial^2 J_X}{\partial x^2}\frac{\Delta x^2}{3!} + \cdots\right) dy\,dz}_{FRONT}$$
$$- \underset{\Delta z\, \Delta y}{\int \int}\underbrace{\left(J_X - \frac{\partial J_X}{\partial x}\frac{\Delta x}{2!} + \frac{\partial^2 J_X}{\partial x^2}\frac{\Delta x^2}{3!} + \cdots\right) dy\,dz}_{BACK} \qquad (1.58)$$

where integration over the y and z ranges of the surfaces is indicated by the subscripts of Δy and Δz, respectively. All the even-powers of Δx terms of the integrals cancel leaving only the odd-powers in the form $(\Delta x)^n$ where $n = 1, 3, \ldots$. Making this simplification and neglecting all but $n = 1$ terms, we obtain

$$\Delta I_F + \Delta I_{BACK} = \int_{z'=z-\frac{\Delta z}{2}}^{z+\frac{\Delta z}{2}} \int_{y'=y-\frac{\Delta y}{2}}^{y+\frac{\Delta y}{2}} \left(\frac{\partial J_X}{\partial x}\Delta x + \cdots\right) dy'\,dz' \qquad (1.59)$$

where the primes indicate the variables of integration about the center point. As the dimensions of the box become vanishingly small, the higher order terms, i.e., $n \geq 3$, become negligible since $\Delta x \gg \Delta x^n$. Furthermore, as Δy and Δz approach 0 and the range of integration shrinks, the integrand is essentially constant over the area of integration. Consequently, the integral can be approximated as the product of the integrand and the area of the integration

$$\Delta I_F + \Delta I_{\text{BACK}} \approx \frac{\partial J_X}{\partial x} \Delta x \Delta y \Delta z = \frac{\partial J_X}{\partial x} \Delta \forall \qquad (1.60)$$

where $\Delta \forall$ is the volume of the rectangular box.

What a simple expression! But what does it say, and, more importantly, what does it mean? The sum of the currents out of the surfaces perpendicular to the x-axis depends upon the derivative of J_X with respect to x and upon the volume of the vanishingly small rectangular box. When J_X is constant, i.e., $\partial J_X / \partial x = 0$, the x-component of the current out of the volume is zero, sort of a one-dimensional KCL. Under this condition, the current that exits one of the surfaces is equal to the current which enters the other. When J_X is a function of x, i.e., $\partial J_X / \partial x \neq 0$, the sum of the two currents is nonzero. When $\partial J_X / \partial x > 0$, the sum of the x-component of the currents is out of the box; when $\partial J_X / \partial x < 0$, the sum is into the box. Since the two surfaces are of equal area, the x-directed current can be nonzero only if J_X varies with respect to x.

There is nothing unique about the x-axis, so the y- and z-directed currents show similar behavior as

$$\Delta I_R + \Delta I_L \approx \frac{\partial J_Y}{\partial y} \Delta x \Delta y \Delta z = \frac{\partial J_Y}{\partial y} \Delta v \qquad (1.61)$$

and

$$\Delta I_T + \Delta I_{\text{BOT}} \approx \frac{\partial J_Z}{\partial z} \Delta x \Delta y \Delta z = \frac{\partial J_Z}{\partial z} \Delta \forall. \qquad (1.62)$$

The total current leaving the box is the sum of Eqs. (1.60), (1.61), and (1.62), and is expressed as

$$\Delta I = \left(\frac{\partial J_X}{\partial x} + \frac{\partial J_Y}{\partial y} + \frac{\partial J_Z}{\partial z} \right) \Delta \forall. \qquad (1.63)$$

The quantity in the parenthesis occurs so frequently in all areas of physics that it has been given the name of "divergence of **J**" or div**J**. With this definition, Eq. (1.63) becomes

$$\Delta I = (\text{div}\mathbf{J}) \Delta \forall. \qquad (1.64)$$

Equation (1.64) is the basis of a physical definition of the divergence as

$$\text{div}\mathbf{J} = \lim_{\Delta \forall \to 0} \left(\frac{\Delta I}{\Delta \forall} \right) = \frac{\partial J_X}{\partial x} + \frac{\partial J_Y}{\partial y} + \frac{\partial J_Z}{\partial z}. \qquad (1.65)$$

As the rectangular box becomes vanishingly small, the ratio of the current that emanates from the box to the volume of the box is defined as the *divergence* of the flux density. Since the

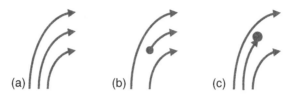

FIGURE 1.15: Flux lines for (a) divF $= 0$, (b) divF > 0, and (c) divF < 0.

box becomes vanishingly small, divJ describes this property of the flux density at a point, the center of the box. At points with zero divergence, flux lines are continuous—none beginning nor ending—so that the amount of flux entering the small box is the same as that leaving, see Fig. 1.15(a). The divergence is nonzero only when there is a net current out of (or into) the box. This requires that an excess of flux lines leave (or enter) the box which is possible only when lines of flux begin (or end) within the box. In other words, the flux "diverges" from such a point. Consequently, there must be a *source* of flux at points where the divergence is positive, see Fig. 1.15(b), and a *sink* at points where it is negative, see Fig. 1.15(c). For electric currents, moving charge is the source (or sink) of the flux. Charge conservation requires that the net motion of charge out of (or into) a point is accompanied by a depletion (or accumulation) of total charge at that point. Such a point is called a source (or sink) of current. divJ $\neq 0$ indicates the depletion or accumulation of charge at a point.

Up to this point, we have assumed that the divergence exists and is finite. We have ignored the fact that the derivatives that define the divergence are infinite whenever the flux density is discontinuous. Though discontinuous flux density can occur in problems concocted by perverse ECE faculty, these situations rarely occur in nature. The flux density may change very rapidly in some regions, but on a small enough scale the change is finite.

Measurements to confirm the divergence of a flux density cannot be made since instrumentation is not differentially small, but always occupies some finite region. However, if incremental rather than differential dimensions are considered, the divergence can be approximated as

$$\text{divJ} \approx \frac{\Delta J_X}{\Delta x} + \frac{\Delta J_Y}{\Delta y} + \frac{\Delta J_Z}{\Delta z}. \tag{1.66}$$

In addition, this approximate form is useful for numeric computations as long as the spatial increments are not too large. If in doubt about the increment size, halve it and compare the results with the original.

A useful approximation of an integral over a surface comes from Eq. (1.64) as well. The net flux out can be estimated in terms of the divergence as

$$\Psi_{OUT} = \oiint_S \mathbf{F} \cdot \mathbf{ds} \approx (\text{div}\mathbf{F})\Delta\forall \qquad (1.67)$$

as long as the divergence exists and is finite.

A mathematically compact notation for the divergence comes from the recognition that the derivatives of the nabla or del operator are present in the divergence. A careful look at Eqs. (1.16) and (1.65) reveals that the scalar derivative associated with a unit vector of the del operator acts upon the corresponding component of the flux density. For example, the x-component of del, $\partial/\partial x$, operates upon J_X. Recall that the dot product of two vectors is the scalar multiplication of their respective components. Formal application of this principle treats ∇ as a vector with an x-component of $\partial/\partial x$ so that when it is dotted with $\mathbf{J}$ the scalar product of the two x-components becomes $\partial J_X/\partial x$. Obviously, the dot product $\mathbf{J} \cdot \nabla$ does not have the same meaning; the dot product is not necessarily commutative with operators. This approach enables us to express the divergence as the dot product of del and flux density as

$$\text{div}\mathbf{J} = \nabla \cdot \mathbf{J} = \frac{\partial J_X}{\partial x} + \frac{\partial J_Y}{\partial y} + \frac{\partial J_Z}{\partial z} \qquad (1.68)$$

This is by far the most common notation for the divergence.

The divergence can be expressed in cylindrical and spherical coordinates though its form is different than for Cartesian coordinates, see Appendix C. Alternatively, the divergence in these coordinate systems can be obtained by applying the derivation procedures used for a rectangular box to a curvilinear box in cylindrical or spherical coordinates.

Example 1.19-1. A flux density is given by $\mathbf{F} = (-x^2 + 4x)\mathbf{a}_X - 3y^2z^2\mathbf{a}_Y - (2y^3z^2 - z)\mathbf{a}_Z$. Calculate div$\mathbf{F}$. Application of Eq. (1.68) leads to $\nabla \cdot \mathbf{F} = 4 - 2x - 6yz^2 - 4y^3z + 1 = 5 - 2x - 6yz^2 - 4y^3z$.

Example 1.19-2. A current density in a resistor is purported to be given by $\mathbf{J} = x\mathbf{a}_X - \mathbf{a}_Y$ A/m^2. Is this possible? Within a resistor we know that div$\mathbf{J} = 0$ since all the current that enters leaves. Application of Eq. (1.68) shows that $\nabla \cdot \mathbf{J} = 1 \neq 0$; the current flux diverges. The positive value indicates that positive charge is emanating from every point so that the rate of change of charge within the resistor is negative. This is not a resistor!

Example 1.19-3. Repeat Example 1.19-2 with a current density of $\mathbf{J} = x\mathbf{a}_X - y\mathbf{a}_Y$ A/m^2. Equation (1.68) reveals $\nabla \cdot \mathbf{J} = 1 - 1 = 0$. Since the current density is divergenceless, this could be the current within a resistor.

1.20   FLUX TUBES

Throughout a region where $\nabla \cdot \mathbf{F} = 0$ there are no sources or sinks. Flux lines do not begin or end within the region, but are continuous. This is the case for current flow in resistors where $\nabla \cdot \mathbf{J} = 0$ everywhere. Since there is no charge accumulation anywhere within the resistor, $\mathbf{J}$ flux lines begin and end outside the resistive material. Moreover, since $\mathbf{J}$ lines coincide with $\mathbf{E}$ lines in isotropic materials, they are always perpendicular to equipotential surfaces. Consequently, $\mathbf{J}$ lines never cross themselves. This suggests a model in which current flows within tubes that extend from one electrode to the other, never crossing from one tube to another. This concept is used often to sketch flux lines where each line represents a *tube of flux* with the flux in all tubes equal. In regions of high flux density, the lines are packed together closely; where the flux density is low, the lines are relatively far apart.

For tubular resistors considered earlier, we found the current flow is in a straight line. But, how does $\mathbf{J}$ behave when the resistor has a cross-sectional area that varies along the axis of the resistor or when the resistor bends around a corner? We are not quite prepared to make an exact calculation, but we can gain an approximate idea of what occurs. Consider the "two-dimensional" resistor with a nonuniform cross section as shown in Fig. 1.16. Such a resistor is a model for a three-dimensional resistor with constant depth into the plane of the paper. With no variations in depth, the fields in every y-plane are the same, e.g., $\partial J_Y / \partial y = 0$. Hence, the fields in any of these two-dimensional planes represent the fields at any depth. Though the resistor cross-section varies along the axis of the resistor, we can describe the fields intuitively. Because charge cannot accumulate within a resistor, the current is the same at every cross section. Therefore, the current density varies inversely with cross-sectional area. For larger cross sections, the current density is smaller; for smaller cross sections, the current density is larger. In sections where the resistor's height is reduced, the current is "squeezed" in the x-direction resulting in an increased current density. When the height increases, the current "expands" to decrease the current density. The divergence offers us another way to look at this phenomenon.

For the resistor shown in Fig. 1.16, $\partial J_Y / \partial y = 0$ (where y is directed perpendicular to the plane of the paper) so that $\nabla \cdot \mathbf{J} = 0$ becomes

$$\frac{\partial J_Z}{\partial z} = -\frac{\partial J_X}{\partial x}. \qquad (1.69)$$

A change in the transverse (to the horizontal axis) current density, J_X (where x is directed upward), is accompanied by a change in the axial current density, J_Z. Though the lines of current shown in Fig. 1.16 are only approximate, they indicate the noncrossing nature of current flux tubes. On the left-hand side of the resistor, $\partial J_X / \partial x < 0$, so that Eq. (1.69) shows $\partial J_Z / \partial z > 0$ just as expected; as the cross section decreases, the axial current density must

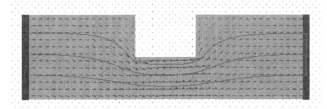

FIGURE 1.16: Current in a nonuniform resistor.

increase. The opposite occurs on the right-hand side where $\partial J_X/\partial x > 0$ and $\partial J_Z/\partial z < 0$ due to the increasing cross section.

Increases in current density in one direction can only occur if there are compensating decreases in current density in other directions. This forms a sort of conservation of changes in flux density that occurs in regions where the divergence vanishes, i.e., regions where the charge is unchanging. This concept gives us insight as to what to expect for current flow in various shaped resistors even without the mathematical details of exact solutions.

1.21 DIVERGENCE THEOREM

Karl Frederich Gauss, a superb mathematical physicist of the 19th century, used the ideas of the last section to prove what is known as the *divergence theorem*. It is expressed as

$$\Psi = \oiint_S \mathbf{F} \cdot \mathbf{ds} = \iiint_\forall \nabla \cdot \mathbf{F} dv \qquad (1.70)$$

The flux emanating from a closed surface S is equal to the integral of the divergence of the flux density throughout the volume $\forall$ enclosed by the surface S. Of course this assumes that $\nabla \cdot \mathbf{F}$ exists and is finite everywhere within the volume $\forall$. Any good mathematician can prove this true with rigorous arguments, but a simpler, plausible argument establishes the validity of this law. Subdivide the volume $\forall$ into incrementally small subvolumes. Since flux emanates from an incremental volume only if $\nabla \cdot \mathbf{F} \neq 0$, all regions for which $\nabla \cdot \mathbf{F} = 0$ are discarded in calculations of the flux emanating from S. The flux from the remaining regions is summed up or integrated to obtain the flux from S.

The divergence theorem provides another, often simpler, method for calculating the flux from a closed surface. But, it has a couple of limitations. Firstly, it cannot be used when the divergence is undefined or infinite. For example, the flux density of a point source is of the form $1/r^2$ that is undefined at the origin. The flux emanating from a closed surface can be readily calculated from the surface integral, but it cannot be determined using the volume integral.

Secondly, the divergence theorem does not apply to open surfaces since there is no enclosed volume throughout which to integrate the divergence.

Example 1.21-1. Calculate the current emanating from the unit sphere for the isotropic current density given by $\mathbf{J} = (1/r^2)\mathbf{a}_r$ A/m². Calculation of the divergence (see Appendix C for the form of the divergence in spherical coordinates) leads to $\nabla \cdot \mathbf{J} = (1/r^2)\partial(r^2/r^2)/\partial r = (1/r^2)\partial(1)/\partial r$ which vanishes except at the origin where it is undefined due to the $1/r^2$ term. Because the $\nabla \cdot \mathbf{J}$ is undefined within the volume, we cannot apply the divergence theorem to this case. We must calculate the current by the surface integral as in Example 1.18-1.

Example 1.21-2. Reconsider the flux density of Example 1.18-3 with $\mathbf{F} = (\rho^2 \cos\phi)\mathbf{a}_\rho + 5\rho z\sin\phi\mathbf{a}_\phi + 3\rho\mathbf{a}_z$ w/m². The divergence of $\mathbf{F}$ is $\nabla \cdot \mathbf{F} = 3\rho\cos\phi + 5z\cos\phi$. Integrating this throughout the volume leads to

$$\Psi = \int_{z=-1}^{3} \int_{\phi=0}^{\pi/4} \int_{\rho=0}^{2} (3\rho + 5z)\cos\phi\,\rho\,d\rho\,d\phi\,dz$$

$$= \left(\frac{1}{\sqrt{2}}\right)[8(4) + 5(4)(2)] = 50.912\,\text{w}$$

as we obtained from the surface integral.

Example 1.21-3. Calculate the current emanating from the unit sphere due to flux density $\mathbf{J} = x\mathbf{a}_X + y\mathbf{a}_Y + z\mathbf{a}_Z$ A/m². This leads to a very complicated surface integral as we convert the Cartesian coordinates and unit vectors to the required spherical coordinates of the surface. But, if we use the divergence theorem, the work is nearly trivial since $\nabla \cdot \mathbf{J} = 3$. Consequently, the integrand is a constant so that the integral becomes the volume of the unit sphere, $4\pi/3 \times \nabla \cdot \mathbf{J}$. The current out of the sphere is $I = 4\pi\,A$.

1.22 CHARGE CONSERVATION REVISITED

The concept of point charges is very familiar to us. For circuit analysis, a quite accurate model for current within a wire is the motion of lumped units of charge called electrons. In spite of their conceptual or mathematical convenience, point charges do not exist. Instead all charge is distributed throughout a volume. For some distributions, the volume is exceedingly small so that it is quite accurate and convenient to model it as a point charge. This suggests that we take a different view of electric charge. Rather than focusing on the total charge within a finite volume, we can consider the spatial distribution of the charge and to express its volume charge density at each point. *Volume charge density* is defined as the ratio of total charge contained within a region to the volume of the region as the region shrinks uniformly to zero, i.e.,

$$\rho_v = \lim_{\Delta \forall \to 0} \frac{\Delta Q}{\Delta \forall} = \frac{dQ}{dv} \; \left[\text{C/m}^3 \right]. \tag{1.71}$$

Volume charge density expressed as Coulombs/cubic meter is analogous to the description of mass by kilograms/cubic meter. An incremental volume contains an incremental amount of charge expressed as

$$\Delta Q = \rho_V \Delta \forall. \tag{1.72}$$

The charge contained within a finite region can be obtained by summing the contributions of each incremental volume as

$$Q = \sum_{i=1}^{N} \Delta Q_i = \sum_{i=1}^{N} \rho_{vi} \Delta v_i \tag{1.73}$$

where ρ_{vi} is the charge density within the ith incremental volume Δv_i. Of course this concept can be extended to differential calculus as

$$Q = \iiint_\forall \rho_v dv. \tag{1.74}$$

The triple integral is the usual notation for three-dimensional scalar integration.

When charge is spread so thinly over a surface that it can be approximated as having zero thickness, it is defined as a *surface charge density*, ρ_S, in Coulombs/square meter,

$$\rho_S = \lim_{\Delta a \to 0} \frac{\Delta Q}{\Delta a} = \frac{dQ}{da} \; \left[\text{C/m}^2 \right]. \tag{1.75}$$

The total charge on a surface is calculated by summing the contribution of each incremental area as

$$Q = \sum_{i=1}^{N} \Delta Q_i = \sum_{i=1}^{N} \rho_{Si} \Delta a_i \tag{1.76}$$

where ρ_{Si} is the surface charge density of the ith surface element, Δa_i. In integral form, the charge is calculated by

$$Q = \iint_S \rho_S da. \tag{1.77}$$

The double integral is a scalar integral over the surface, not to be confused with a flux integral. Evaluation can be accomplished by the usual techniques of multiple scalar integrals.

Often charge is arranged along a linear axis, as on a wire, with a diameter that is small compared to its length. This distribution of charge is known as *line charge density* and defined as

$$\rho_L = \lim_{\Delta l \to 0} \frac{\Delta Q}{\Delta l} = \frac{dQ}{dl} \; [C/m] . \tag{1.78}$$

The total charge on a line is calculated by summing the contribution of each incremental length as

$$Q = \sum_{i=1}^{N} \Delta Q_i = \sum_{i=1}^{N} \rho_{Li} \Delta l_i \tag{1.79}$$

where ρ_{Li} is the line charge density on the ith incremental line element, Δl_i. In integral form, this becomes

$$Q = \int_L \rho_L dl . \tag{1.80}$$

Of course, *point charges*, denoted as q, are often a good approximation for highly localized charge distributions such as electrons. They have units of Coulombs.

In reality, all charges are distributed throughout a volume modeled by ρ_V. The other distributions are approximations which model several important special cases. If one dimension of this volume is vanishingly small relative to the other two, then a good model is surface charge density, ρ_S. If two dimensions are relatively small, the line charge density is an appropriate model, ρ_L. Finally, if all three dimensions are vanishingly small as well, then the point charge model is appropriate, q. The term "relatively small" describes the relationship of the charge distribution compared to the distance between the observer and the distribution. Figure 1.17 shows the several models for charge distributions. As an engineer you must choose the model most appropriate for the problem at hand.

For simplicity, these charge distributions are often assumed to be uniformly distributed. No matter how convenient this approximation may be, it is rarely true. ρ_V, ρ_S, and ρ_L usually are functions of space.

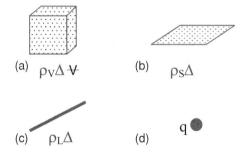

FIGURE 1.17: Charge distribution models: (a) ρ_V, (b) ρ_S, (c) ρ_L, and (d) q.

Finally, let's apply the divergence theorem to Eq. (1.47) as

$$I = \oiint_S \mathbf{J} \cdot \mathbf{ds} = \iiint_\forall \nabla \cdot \mathbf{J} dv = -\frac{dQ}{dt}. \qquad (1.81)$$

If we restrict this development to regions with only volume charge distributions, then we can represent Q by Eq. (1.74) and the right-most equality can be expressed as

$$\iiint_\forall \nabla \cdot \mathbf{J} dv = -\frac{dQ}{dt} = -\frac{d}{dt}\left(\iiint_\forall \rho_V dv\right). \qquad (1.82)$$

For a fixed volume $\forall$, the time derivative can be taken inside the integral and replaced by a partial derivative operating on ρ_V as

$$\iiint_\forall \nabla \cdot \mathbf{J} dv = -\iiint_\forall \frac{\partial \rho_V}{\partial t} dv. \qquad (1.83)$$

Since both integrals are over the same volume, these two integrals can be combined into one integral as

$$\iiint_\forall \left(\nabla \cdot \mathbf{J} + \frac{\partial \rho_V}{\partial t}\right) dv = 0 \qquad (1.84)$$

and since it holds for all points within the volume, the integrand must vanish as

$$\nabla \cdot \mathbf{J} = -\frac{\partial \rho_V}{\partial t}. \qquad (1.85)$$

This is the differential form of charge conservation. $\mathbf{J}$ lines begin or end at a point, i.e., $\nabla \cdot \mathbf{J} \neq 0$, only if there is a change in the charge density at the point. Of course, if ρ_V is constant, the $\mathbf{J}$ lines are continuous. In the case of resistors, $\rho_V = 0$ and Eq. (1.85) takes the form of

$$\nabla \cdot \mathbf{J} = 0. \qquad (1.86)$$

Unfortunately, the partial derivatives of $\nabla \cdot \mathbf{J}$ become infinite when surface, line, or point charge distributions are present and the differential form of charge conservation fails. Nevertheless, this form is applicable widely enough to make it useful.

Example 1.22-1. A line charge density is $\rho_L = |x - L/2|$ C/m in the range $-L/2 \leq x \leq L/2$. Calculate the total charge in this distribution. Applying Eq. (1.80) and using the symmetry of the distribution, we find

$$Q = \int_{-L/2}^{L/2} \rho_L dx = \int_{-L/2}^{L/2} |x - L/2| dx$$

$$= 2 \int_{0}^{L/2} (L/2 - x) dx = 0.25 L^2 \ C.$$

Example 1.22-2.   A sphere of radius $r = a$ encloses part of a volume charge density of $\rho_V = 10 \exp[-r^3]$ C/m^3. What is the radius of the sphere if the total charge enclosed is 10 C? The total charge within the sphere is expressed as

$$Q = 10 = \int_{r=0}^{a} \int_{\theta=0}^{\pi} \int_{\phi=0}^{2\pi} 10 e^{-r^3} r^2 \sin\theta \, dr \, d\theta \, d\phi$$

$$= 10 \left(\int_{r=0}^{a} e^{-r^3} r^2 dr \right) \left(\int_{\theta=0}^{\pi} \sin\theta \, d\theta \right) \left(\int_{\phi=0}^{2\pi} d\phi \right)$$

$$= \left(\frac{10}{3} \right) (1 - e^{-a^3})(2) 2\pi = \frac{40\pi}{3} (1 - e^{-a^3}).$$

Using a numerical solution with MAPLE or similar software, we obtain $a = 0.65$ m.

Example 1.22-3.   What is the rate of change of the charge density in Example 1.21-1? The current density is $\mathbf{J} = x\mathbf{a}_X + y\mathbf{a}_Y + z\mathbf{a}_Z$ A/m^2, so that $\nabla \cdot \mathbf{J} = 3$. Therefore, $\partial \rho_V / \partial t = -3$ C/m^3/s; the volume charge density is decreasing at a constant rate throughout the region where the current density exists.

1.23   OHM'S LAW

A concise summary of our knowledge so far predicts the presence of an electric field intensity within a resistor which is proportional to the voltage drop across it and an accompanying current flux density that is proportional to the current within the resistor. From circuits we know that the voltage drop is proportional to the current, $V_R = I_R R$, so the current flux density must be related to the electric field intensity. That is the focus of this section.

We have established that for an axial resistor $\mathbf{E}$ points from more positive to less positive voltage described by Eq. (1.10) as $\mathbf{E} = -(\Delta V/\Delta z)\mathbf{a}_Z$. From Eq. (1.37), $\mathbf{J} = -(\Delta I/\Delta a)\mathbf{a}_Z$, we found that $\mathbf{J}$ flows in the same direction as $\mathbf{E}$. Combining these with Eq. (1.2), $R = (L/A\sigma)$, we obtain

$$\mathbf{E} = -\frac{\Delta V}{\Delta z}\mathbf{a}_Z = -\frac{V}{L}\mathbf{a}_Z = -\frac{1}{\sigma}\left(\frac{I}{A}\right)\mathbf{a}_Z$$
$$= -\frac{1}{\sigma}\left(\frac{\Delta I}{\Delta a}\right)\mathbf{a}_Z = \frac{\mathbf{J}}{\sigma}, \qquad (1.87)$$

or in the more common form

$$\mathbf{J} = \sigma\mathbf{E}, \qquad (1.88)$$

which is known as the *point form of Ohm's law*. The current density is proportional to the electric field intensity in a manner similar to the proportional relationship between current and voltage. Resistance establishes the proportionality between the voltage and current of circuits. From physics we know that resistance depends upon two parameters of the resistor—its conductivity and its geometry. However, the field parameters show a striking difference. The electric field and the current density are related by a single parameter—the conductivity of the material. Experimental data show that Eq. (1.88) is valid for all materials, independent of shape or size. This simpler relationship suggests that the field representation of Ohm's law is a more fundamental form than in circuits. As a matter of fact, expressions for resistance of more complex resistor shapes are much more complicated. In the following sections, we will use these simpler field concepts to determine the resistance for any geometry.

1.24 CONDUCTIVITY

Current flow in solid materials is characterized by the parameter conductivity denoted by the symbol σ with units of $(\Omega m)^{-1}$ known as Siemens/meter, in the SI system. Alternatively, resistivity, denoted by the symbol $\rho = 1/\sigma$ and with units of Ωm, is used. We will use σ throughout this text to avoid confusion with charge density.

Charge carriers, most often electrons, within conductive materials experience the Coulomb force, $\mathbf{F} = q\mathbf{E}$, due to an applied electric field. Those carriers that are free to move are accelerated by this field, but experience frequent collisions with the atoms of the lattice structure. The carriers transfer some of their kinetic energy to the atoms, rebound, and are accelerated by the Coulomb force again. The energy transferred to the atoms causes them to vibrate within the lattice and is observed as heating of the material. This is called ohmic heating and is related to the electric power supplied by the applied field; we will look at this later. The more electrons which are moved and the easier they are moved the greater the value of conductivity. Materials are classified by their conductivity.

There are three broad categories of conductive materials—conductors, semiconductors, and insulators—depending upon the value of conductivity. Conductors are most often metals such as gold, silver, copper, aluminum, or iron. A virtual "sea" of electrons that occupy the conduction band are rather loosely bound to the atoms so that one or more electrons are available for each atom. These electrons are readily moved by applied electric fields. Moreover, there are typically 10^{28} atoms/m^3 so that a huge number of electrons are available. These conditions result in conductivities of metals in the order of 5×10^7 S/m.

Semiconductors, used in many solid-state devices, are "half" conductors as the name implies. The number of charge carriers in the conduction band is controlled by doping levels and is very temperature dependent. However, the number of carriers is significantly less than that of metals. Conductivities are typically in the order of $10^3 - 10^5$ S/m.

The electrons in insulators are bound extremely tightly to the atoms; there are no conduction electrons. The dielectric materials used in capacitors are a form of insulator. A very large field is required to release electrons from their central atoms. This process, usually accompanied by an arc, is known as *dielectric breakdown*. Usually the very few charge carriers that exist are due to internal or surface imperfections. Insulators have conductivities in the order of 10^{-9} to 10^{-17} S/m.

Resistors are commonly made of graphite or a similar material with conductivity in the order of 7×10^4 S/m. Power resistors usually consist of several turns of a high resistance wire such as a nichrome wire with a conductivity of 10^6 S/m. The conductivity of fresh water is in the order of 10^{-3} S/m while the dissolved salt ions in sea water increase its conductivity to about 1 S/m. Soil conductivity varies greatly depending upon the water content with an average value of 10^{-4} S/m. Air is often treated as a vacuum—an ideal insulator—with zero conductivity. Though a quite satisfactory insulator when the relative humidity is low, it has significant conductivity in humid weather.

We will consider two ideal cases—*perfect electric insulators* with $\sigma = 0$ and *perfect electric conductors* with $\sigma = \infty$. Perfect electric conductors are commonly known as PECs. A perfect insulator can have no current flow since $\sigma = 0$. A PEC can have neither a finite electric field nor a finite voltage drop. This is because a finite electric field when multiplied by $\sigma = \infty$ would result in an infinite current density, a physically unrealizable condition. In circuits, we assume that the interconnecting wires of a circuit have zero resistance. This is the same as assuming that they are made of PEC material. Since there is no voltage drop in PEC material, it is lossless as well. More on this will be discussed later.

For more detailed descriptions of charges and current flow in materials, consult a textbook in solid-state physics. The descriptions above are purposely kept as simple as possible so as not to detract from our study of electromagnetics. Tables of material properties including conductivity may be found in Appendix D.

1.25 BOUNDARY CONDITIONS

The concepts developed to this point have dealt with fields in a continuous medium with no boundaries present. However, boundaries between two different materials are usually involved in resistors and other electromagnetic elements. So we need to understand the behavior of fields at boundaries. A typical and useful strategy is to decompose the field vectors into those components that are perpendicular to and those that are parallel to the interface between two different materials. Figure 1.18 illustrates the components of typical **E** and **J** fields near a boundary.

The conservative nature of electric fields, i.e., they satisfy KVL around any closed loop, governs the behavior of **E** at a boundary. This is accomplished by applying Eq. (1.25) to the closed path in Fig. 1.18,

$$\oint_{L_1-L_2} \mathbf{E} \cdot \mathbf{dl} = \int_{L_1} \mathbf{E} \cdot \mathbf{dl} - \int_{L_2} \mathbf{E} \cdot \mathbf{dl} = 0, \qquad (1.89)$$

which can be rewritten as

$$\int_{L_1} \mathbf{E} \cdot \mathbf{dl} = \int_{L_2} \mathbf{E} \cdot \mathbf{dl} \qquad (1.90)$$

where L_1 is the directed segment of the path in region 1 and L_2 in region 2.

Since we are interested in obtaining the conditions that **E** must satisfy at every point on the boundary, these are known as *boundary conditions*. Behavior at a point is obtained if the length and width of the paths L_1 and L_2 are shrunk to zero in which case the integrals can be approximated by

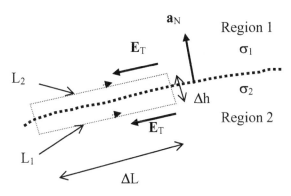

FIGURE 1.18: Electric field boundary conditions.

$$\int_{L_1} \mathbf{E} \cdot \mathbf{dl} \approx -E_{T1}\Delta L + (E_{N1L} - E_{N1R})\left(\frac{\Delta h}{2}\right)$$

$$\int_{L_2} \mathbf{E} \cdot \mathbf{dl} \approx -E_{T2}\Delta L + (E_{N2R} - E_{N2L})\left(\frac{\Delta h}{2}\right) \tag{1.91}$$

where E_{Ti} is the tangential field at the boundary in the ith region and E_{NiR} and E_{NiL} are the normal components in the ith region at the right and left ends of the path, respectively. Substituting these results into Eq. (1.90), we obtain

$$(E_{N1L} + E_{N2L} - E_{N1R} - E_{N2R})\left(\frac{\Delta h}{2}\right)$$

$$- (E_{T1} - E_{T2})\Delta L = 0. \tag{1.92}$$

If $\Delta h \ll \Delta L \to 0$, then the first term in parentheses contributes insignificantly compared to the second as long as all of the normal components are finite. Only the last term remains so that

$$E_{T1} = E_{T2} \tag{1.93}$$

or in the vector form

$$\mathbf{a}_N \times (\mathbf{E}_1 - \mathbf{E}_2) = 0 \tag{1.94}$$

as the boundary condition for electric fields. The unit normal to the interface, $\mathbf{a}_N$, points from region 2 to region 1. Only the tangential components of $\mathbf{E}$ are affected by the boundary. The normal components are governed by other conditions. Note that the KVL interpretation of Eq. (1.90) is that *the voltage drops along the two sides of the boundary are equal*, i.e., $E_{T1}\Delta L_1 = E_{T2}\Delta L_2$.

Now, we must determine the behavior of current density at boundaries, see Fig. 1.19.

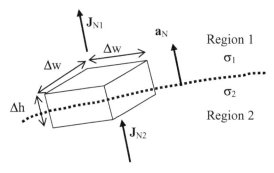

FIGURE 1.19:   Current density boundary conditions.

According to Eq. (1.48), current density satisfies KCL since charge does not accumulate within the resistive material

$$I_{OUT} = \oiint_S \mathbf{J} \cdot \mathbf{ds} = 0. \tag{1.95}$$

Applying this to the short, square box of Figure 1.18, we obtain

$$\oiint_S \mathbf{J} \cdot \mathbf{ds} = \iint_{SIDES} \mathbf{J} \cdot \mathbf{ds} + \iint_{TOP} \mathbf{J} \cdot \mathbf{ds} + \iint_{BOTTOM} \mathbf{J} \cdot \mathbf{ds} = 0. \tag{1.96}$$

According to Eq. (1.48), current density satisfies KCL since charge does not accumulate within the resistive material

$$I_{OUT} = \oiint_S \mathbf{J} \cdot \mathbf{ds} = 0. \tag{1.97}$$

Applying this to the short, square box, we obtain

$$\oiint_S \mathbf{J} \cdot \mathbf{ds} = \iint_{SIDES} \mathbf{J} \cdot \mathbf{ds} + \iint_{TOP} \mathbf{J} \cdot \mathbf{ds} + \iint_{BOTTOM} \mathbf{J} \cdot \mathbf{ds} = 0. \tag{1.98}$$

As with line integrals, the boundary conditions are to apply at a point so we require that $\Delta h \ll \Delta w \to 0$ and Eq. (1.98) becomes

$$\oiint_S \mathbf{J} \cdot \mathbf{ds} \approx (\mathbf{J}_{TOP} - \mathbf{J}_{BOTTOM}) \cdot \mathbf{a}_N \Delta w \Delta w$$

$$+ \sum_{SIDES} \mathbf{J}_{SIDES} \cdot \mathbf{a}_{SIDES} \Delta h \Delta w. \tag{1.99}$$

The second term is negligible compared to the first term as long as $\mathbf{J}_{SIDES}$ remains finite. The only contributions come from the current that flows through the top and bottom of the box and since no charge accumulates within the resistor they must be equal. Rewriting the current densities in terms of normal and tangential, we find

$$J_{N1} = J_{N2} \tag{1.100}$$

or in vector form

$$\mathbf{a}_N \cdot (\mathbf{J}_1 - \mathbf{J}_2) = 0. \tag{1.101}$$

The current density normal to the boundary must be continuous.

The boundary condition of Eq. (1.102) is based upon the condition that there is no change in the charge density on the interface between the two conductive regions. We leave it to advanced electromagnetics courses to determine the boundary conditions when the charge on the boundary is changing with time.

From Eq. (1.93), we see that the tangential components of the electric field intensity must be equal, an analogy with KVL. From Eq. (1.100), we see that the normal components of the current density must be equal, an analogy with KCL. Using the relations $\mathbf{J}_1 = \sigma_1 \mathbf{E}_1$ and $\mathbf{J}_2 = \sigma_2 \mathbf{E}_2$, we can express the relationships for both components of the two fields as

$$E_{T1} = E_{T2} \quad \text{and} \quad \sigma_1 E_{N1} = \sigma_2 E_{N2}$$

$$\frac{J_{T1}}{\sigma_1} = \frac{J_{T2}}{\sigma_2} \quad \text{and} \quad J_{N1} = J_{N2}. \tag{1.103}$$

The tangential components of both fields are governed by the electric field behavior at the boundary and the normal components by the current density.

These results hold important implications for the interfaces between the resistive material and the metal electrodes and the material surrounding the body of the resistor. The electrodes can be modeled as PECs due to their relatively high value of their conductivity compared to the resistive material and insulator. The electric fields and voltage drops in PECs are zero due to their infinite conductivity; PECs form equipotential surfaces. For this reason, the current leaving or entering the resistive material must be perpendicular to the electrodes. The insulator or air surrounding the resistive material is modeled as a perfect insulator with zero conductivity and will not support any current flow. This means that no current can flow out of the resistor into the insulator. Consequently, there can be no normal component of current at the air-resistive material interface. All of the current must be tangential to the conductor–air interface.

These observations suggest that the resistive material "guides" the current flux through the resistor. All the current enters the *flux guide* through the electrode at one end and leaves through the electrode at the other end. None of it leaks out along the path between. In practice, the material surrounding the body of a resistor has nonzero, albeit small, conductivity so that there is a very small leakage current. Nevertheless, the resistive material still functions well as a flux guide. Intuitively, a good flux guide will have a high ratio of flux that flows through the guide to the flux that leaks into the insulator. In spite of a complex geometric dependence, the effectiveness of a material as a flux guide is essentially governed by the ratio of the conductivities of the guide and the surrounding material. This is seen as $J_{T2} = (\sigma_2/\sigma_1)J_{T1}$ from Eq. (1.103). When the resistive material is surrounded by a perfect insulator with $\sigma_2 = 0$, it acts as a perfect flux guide since $\sigma_2/\sigma_1 = 0$. The effectiveness of actual flux guides will be somewhat less since the leakage current will not be zero. A good rule of thumb is that an effective flux guide will have $\sigma_2/\sigma_1 \leq 0.1$. These principles will apply later as we consider other circuit elements with different types of flux densities.

Example 1.25-1. An electric field in region 1 where $\sigma_1 = 1$ S/m has $E_{T1} = 1$ V/m and $E_{N1} = 2$ V/m. Calculate the tangential and normal components of $\mathbf{E}_2$ and $\mathbf{J}_2$ in region 2 where $\sigma_2 = 2$ S/m. Equating tangential components, we obtain $E_{T2} = E_{T1} = 1$ V/m. The normal components of current density are equal so that $E_{N2} = (\sigma_1/\sigma_2)E_{N1} = (1/2)2 = 1$ V/m. Therefore, $J_{T2} = \sigma_2 E_{T2} = 2$ A/m^2 and $J_{N2} = \sigma_2 E_{N2} = 2$ A/m^2.

Example 1.25-2. What is the ratio of tangential flux at the boundary between two conductive materials with $\sigma_1 = 1$ S/m and $\sigma_2 = 8$ S/m? The ratio $\sigma_1/\sigma_2 = 1/8$ suggests that region 2 has eight times more tangential flux than region 1. While not the suggested ratio of 10, region 2 is a reasonably good flux guide and we might well neglect the leakage of current from region 2 into region 1.

1.26 INCREMENTAL RESISTORS

A brief review of our findings so far indicates that

(1) electric field intensity is defined in differential form as $\mathbf{E} = -\nabla V$;

(2) voltage drop between the final and initial point is defined as $V_{\text{FINAL}} - V_{\text{INITIAL}} = -\int_L \mathbf{E} \cdot \mathbf{dl}$;

(3) current density is related to the electric field by $\mathbf{J} = \sigma\mathbf{E}$, where σ is the conductivity of the material;

(4) $\mathbf{J}$ is perpendicular to surfaces of constant voltage V;

(5) resistor current is defined as $I = \iint_S \mathbf{J} \cdot \mathbf{ds}$;

(6) charge does not accumulate at any location within a resistor since $\nabla \cdot \mathbf{J} = 0$.

These are the concepts at our disposal to calculate resistance. Along the way we may find it convenient to calculate fields, voltage, or current.

The basic building block for nearly all resistance calculations is the incremental resistor, see Fig. 1.20. It is formed by a flux tube that carries an incremental current ΔI. The four sides of the flux tube are curved so that they remain parallel to $\mathbf{J}$. The flux tube has an incremental area of Δa. The two ends where current enters or exits the incremental resistor coincide with equipotential surfaces just as if they were composed of PECs. They are separated by an incremental distance Δl and have an incremental potential difference of ΔV

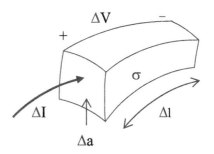

FIGURE 1.20: Incremental resistor.

between them with the current entering the more positive end. Though there are no actual PEC electrodes on either end of the resistor, the current density is perpendicular to the equipotential surface as it is to the PEC electrodes of an actual resistor. Moreover, the current all stays within the flux tube that forms the incremental resistor as it does in an actual resistor. So, let's proceed to calculating the resistance.

The "terminal" current is calculated by integrating **J** over the either end of the incremental resistor. Since it is only an incremental resistance, **J** is essentially constant over the area and the current is given as

$$\Delta I = \iint_S \mathbf{J} \cdot \mathbf{ds} \approx |\mathbf{J}|\Delta a = J\Delta a = \sigma E\Delta a \qquad (1.104)$$

where $\mathbf{J} = \sigma\mathbf{E}$ has been used. The voltage drop across the resistor can be expressed in terms of **E**. Since the resistor is so short, **E** is essentially constant over the length, and the incremental voltage drop is approximated as

$$\Delta V = -\int_L \mathbf{E} \cdot \mathbf{dl} \approx |\mathbf{E}|\Delta l = E\Delta l. \qquad (1.105)$$

Circuit theory provides the definition of resistance as the voltage drop divided by the terminal current which gives the differential resistance of

$$\Delta R = \frac{\Delta V}{\Delta I} = \frac{E\Delta l}{\sigma E\Delta a} = \frac{\Delta l}{\sigma \Delta a}. \qquad (1.106)$$

The incremental resistance is directly proportional to its length and inversely proportional to its cross-sectional area, the same relationship with which we began our discussion of resistors. More importantly, it enables us to use circuit concepts to perform resistance calculations.

Two resistors are considered to be in parallel when they are connected so that they have the same voltage drop. When two incremental resistors terminate on the same equipotentials, see Fig. 1.21, they have the same voltage drop and are considered to be connected in parallel. Instead of the wires used in circuits to establish the equipotentials, the incremental resistors are positioned to terminate on the naturally occurring equipotentials within the resistive material. This allows us to calculate the combined effects

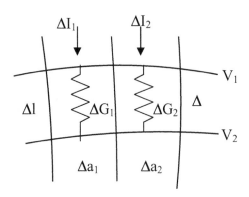

FIGURE 1.21: Parallel incremental resistors.

of the two incremental resistors in parallel, see Fig. 1.21, by adding their conductances

$$\Delta G_{PAR} = \Delta G_1 + \Delta G_2 = \frac{\sigma \Delta a_1}{\Delta l_1} + \frac{\sigma \Delta a_2}{\Delta l_2} \tag{1.107}$$

$$= \frac{\sigma}{\Delta l}(\Delta a_1 + \Delta a_2).$$

The resistors have essentially the same length, Δl, since they are adjacent to each other. A generalization of Eq. (1.107) leads to

$$\Delta G_{PAR} = \frac{\sigma}{\Delta l} \sum_{i=1}^{N} \Delta a_i \tag{1.108}$$

This form is equivalent to the summation of the currents in each of the incremental resistors to obtain the total current in the parallel combination.

Note that Δa is a surface perpendicular to and extending into the plane of the paper. Only one dimension is visible in the two-dimensional sketches such as Fig. 1.21.

Two resistors are considered to be in series when they are connected so that all of the current that leaves one resistor enters the other. The two incremental resistors shown in Fig. 1.22 occupy the same flux tube so that the same current flows through both of them. The combined resistance of the series connection is given by

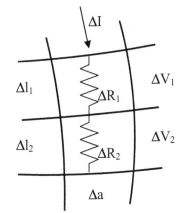

$$\Delta R_{SER} = \Delta R_1 + \Delta R_2 = \frac{\Delta l_1}{\sigma \Delta a_1} + \frac{\Delta l_2}{\sigma \Delta a_2}$$

$$= \frac{(\Delta l_1 + \Delta l_2)}{\sigma \Delta a} \tag{1.109}$$

and in general form for N resistors in series as

FIGURE 1.22: Series incremental resistors.

$$\Delta R_{SER} = \frac{1}{\sigma \Delta a} \sum_{i=1}^{N} \Delta l_i. \tag{1.110}$$

This form is equivalent to the summation of the voltage drops of the incremental resistors.

Resistors of any geometry are composed of a unique circuit of interconnected incremental resistors to which we can apply the concepts of series and parallel circuits. The equivalent resistance for this circuit represents the measured value of resistance between the electrodes. Moreover, this method can be applied when the resistor is composed of different conductivities by aligning flux tubes or equipotentials with the material boundaries.

An alternate perspective emphasizes the relationship between incremental current flux and incremental voltage drop. From Eq. (1.106), the incremental conductance is expressed as

$$\Delta G = \frac{\Delta I}{\Delta V} = \frac{\Delta \Psi_C}{\Delta V} = \frac{\sigma E \Delta a}{E \Delta l} = \frac{\sigma \Delta a}{\Delta l}. \qquad (1.111)$$

The incremental conductance is directly proportional to the conductance of the material. The larger the area of a flux tube, the greater the flux within the incremental element and the greater the conductance. The greater the length of the flux tube, the greater the voltage drop and the less the conductance. We will find that the element values of incremental capacitors and inductors are based on similar relationships between flux and voltage drop. Consequently, their dependence upon the material property (known as constitutive relation) and geometry will be of the same form as Eq. (1.111). The concept embodied by this equation is well worth learning and learning well!

In order to determine the location and orientation of the incremental resistors, we must determine the equipotential surfaces and the directions of current density. We find these by several different methods in the following sections.

1.27 CURVILINEAR SQUARES

For two-dimensional structures where there is no variation in the geometry in the third dimension, voltage and field calculations reduce to a planar problem. This method is essentially a graphical technique based upon the principle that **E** and **J** are perpendicular to equipotentials. You may see this described as *flux mapping* by some authors. By sketching equipotential and flux lines so that they form curvilinear squares, we can find the potential, electric field, and current density at any point within the resistive material. In addition, we can use the methods of the previous section to calculate the total resistance. This seems too easy, but it works because it is based upon the fundamental concept that current flux and equipotential surfaces are perpendicular to each other. It is a graphical solution of the vector equations that govern the behavior of resistors.

Consider the "closed" planar structure of Fig. 1.23 where specified voltages are maintained by an external voltage source on the two PEC electrodes that surround the conductive material. We begin by sketching several flux lines perpendicular to the equipotential boundaries, e.g., the solid flux lines J_1, J_2, and J_3. The exact location of the flux lines is not nearly as important as their perpendicularity to the equipotentials. As a general guideline, at least three or four flux lines emanating from each electrode should give reasonably accurate results.

We only sketch the flux lines a short distance from the electrode since we do not yet know the equipotentials far from the PEC surfaces within the conductive region. Building

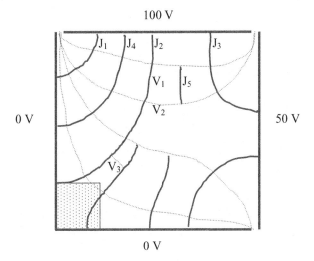

FIGURE 1.23: Voltages and electric field via curvilinear squares.

upon the estimated flux lines, we now sketch an equipotential dashed line V_1 perpendicular to the flux lines to form curvilinear squares. This provides the basis for further extending the flux lines which in turn enables another equipotential dashed line V_2 to be sketched. This process is continued until the entire region between the electrodes is divided into a set of curvilinear squares. It is unlikely that all of our flux and equipotential lines will be mutually perpendicular, so keep an eraser handy and redraw the lines as needed. Repeat the entire process several times until the perpendicularity requirement is satisfied everywhere. Sometimes there will be unusually shaped regions which must be subdivided further as with voltage V_3. Similarly, additional flux lines such as J_4 and J_5 can be added to divide regions into squares. The size of the squares is irrelevant; most importantly, the *flux lines and equipotentials must be perpendicular*. Some regions, such as the shaded region of Fig. 1.23, are not really square. Such regions can be further subdivided, but when they are adjacent to PEC boundaries that form an angle which is less than $180°$ they have a small effect. Moreover, since it is the only nonsquare region, its effect is negligible.

The problem of Fig. 1.23 is a bit unrealistic since all of the potential lines of the conductive region pass through the infinitesimal gaps at the corners between electrodes, creating infinite electric fields there (remember that the electric field intensity is the voltage drop/unit length and the infinitesimal gap has a finite voltage drop). But, we will ignore this since the gaps occupy a very small portion of the region. In an actual problem, the gaps between electrodes are finite and the electric fields are large, but finite. The final result of my sketches is shown in Fig. 1.23. Your results may differ somewhat, but overall, any two sketches will be essentially the same. Remember, this is only a graphical method and relies upon the eye of the solver to make curvilinear squares. The most important aspect of this process is to make sure the flux lines and equipotentials are perpendicular.

This method gives us a good picture of the nature of the current flux and equipotentials throughout the conductive region. However, the voltage of a particular equipotential cannot be determined readily by this method. Of course, all the equipotentials which pass through the upper-left gap of Fig. 1.23 are between 0 and 100 volts, those through the lower-right gap are between 0 and 50 volts, and those through the upper-right gap are between 50 and 100 volts. But which equipotential represents 37 or 81 volts is not known.

In spite of this limitation, we gain a sense of where the electric field is greatest or least from these lines. The magnitude of the electric field at any point is approximated as the voltage difference between the two equipotentials which bound the point; the electric field is perpendicular to the equipotentials and directed from higher to lower voltage. The electric field is greatest in those regions where the equipotentials are most closely spaced, i.e., in the regions of the three electrode gaps. The electric field is least where the lines are far apart as they are in the lower-left corner. Of course, the current density is proportional to the electric field.

Application of the method of curvilinear squares to find equipotentials and current flux as shown in Fig. 1.23 is revealing, but soon we will learn more effective and more powerful techniques. The real utility of this method is in calculating the resistance of arbitrarily-shaped, two-dimensional resistors. We shall demonstrate for the resistor with the cross section shown in Fig. 1.24. The resistor has a conductivity σ and a thickness t into the page. Let's assume an arbitrary voltage drop of 1 V with the upper-left electrode as more positive. As before, the flux lines leave the electrodes perpendicularly. But, we need to look carefully at what happens at the edges of the resistors. Since the conductivity of the dielectric adjacent to the resistor is usually zero, there is no current flow in the air. Consequently, no flux lines leave the edges of the resistor; the flux at the edge is tangential. In addition, this means that equipotentials are perpendicular to the edges. This additional information helps us to sketch the flux lines and equipotentials near the edges. The flux behavior at the edges of resistors surrounded by materials with finite conductivity is much more complicated. But, as long as the conductivity of the surrounding material is much less than that of the resistor, i.e., $\sigma_{SURROUND}/\sigma_{RES} \ll 1$, tangential flux and perpendicular equipotentials are still a reasonable approximation at the edges.

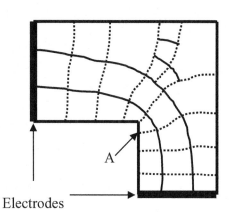

Electrodes

FIGURE 1.24: Calculation of resistance by curvilinear squares.

Now we sketch the flux lines and equipotentials to form curvilinear squares as shown in Fig. 1.24. Each curvilinear square represents an incremental resistor. The resistance between the two electrodes is the equivalent resistance of this series–parallel interconnection.

The conductance of a curvilinear resistor is given as

$$\Delta G = \frac{\Delta I}{\Delta V} = \frac{\Delta \Psi_C}{\Delta V} = \frac{J \Delta a}{E \Delta l} = \frac{\sigma E \Delta a}{E \Delta l} = \frac{\sigma \Delta a}{\Delta l}$$
$$= \frac{\sigma \Delta wt}{\Delta l} \tag{1.112}$$

where the cross-sectional area is given by $\Delta a = \Delta wt$. By making the resistor a curvilinear square, we impose the condition of $\Delta l / \Delta w = 1$, which simplifies the expression for incremental conductance to

$$\Delta G = \sigma t. \tag{1.113}$$

Only the thickness and conductivity of the resistor enter into the incremental resistance of the curvilinear square. The reciprocal of this conductance of a curvilinear square is often denoted by $R_\square$ and is known as the *resistance/square* because it is the resistance of a square regardless of its size.

Example 1.27-1.   Calculate the resistance of the resistor shown in Fig. 1.24 in terms of $1/\sigma t$ using curvilinear squares. In addition, calculate the voltage at point A. Perpendicular equipotential (dashed lines) and flux lines (solid lines) are sketched to form curvilinear squares. Then, the total resistance is obtained readily by application of circuit rules for series- and parallel-connected resistors. The flux tube along the lower-left corner of the resistor is composed of eight squares, incremental resistors in series; so is the center flux tube. The upper-left flux tube is a more complicated connection of resistors, seven in series with the parallel combination of two and the parallel combination of three. Applying series/parallel rules of circuits leads to the total resistance $R = (8 || 8 || (7 + 1/2 + 1/3))/\sigma t = 2.65/\sigma t\ \Omega$. Note that most of the outer tube has roughly the same resistance as the two inner tubes, but with a much greater cross section. Consequently, the current density is much less in this tube, especially near the upper-right corner. The equipotentials are much closer together within the inner flux tube with a greater electric field, especially near the lower-left corner. Finally, the voltage drop across the inner flux tube is divided equally across each of the incremental, curvilinear resistors in the tube. Since there are eight resistors in series and three resistors from point A to ground, the voltage at point A is calculated by voltage division as $V_A = 3/8 = 0.375\ V$.

The method of curvilinear squares is intuitive and powerful, though its accuracy is limited to about 10% due to graphical inaccuracy. Let's consider several other methods that lead to analytic and numeric methods to solve for the voltages and fields within a resistor.

1.28 LAPLACE'S EQUATION

A more mathematical approach combines several previous equations to form a single partial differential equation for the voltage within a resistor. Traditional partial differential equation (PDE) techniques lead to analytic solutions for the voltage. From these analytic solutions, we gain additional insights into the behavior and properties of electric fields.

We will use three equations from our previous work in developing this PDE. Firstly, we found that the electric field is the negative gradient of the voltage, i.e.,

$$\mathbf{E} = -\nabla V. \tag{1.114}$$

Secondly, there is no accumulation of charge throughout the conductive region, i.e.,

$$\nabla \cdot \mathbf{J} = 0. \tag{1.115}$$

Finally, Ohm's law in point form is given by

$$\mathbf{J} = \sigma \mathbf{E}. \tag{1.116}$$

Substituting Eq. (1.116) into Eq. (1.115) gives us

$$\nabla \cdot (\sigma \mathbf{E}) = \frac{\partial^2 (\sigma E_X)}{\partial x^2} + \frac{\partial^2 (\sigma E_Y)}{\partial y^2} + \frac{\partial^2 (\sigma E_Z)}{\partial z^2} = 0. \tag{1.117}$$

The derivatives in the RHS depend upon the properties of the conductivity of the resistive material. Within most resistors, the conductivity does not vary from point to point; such material is called *uniform* or *homogeneous*. Materials in which the conductivity varies from point to point are called *nonuniform* or *inhomogeneous*. Another property of most conductors is that the conductivity is the same regardless of the direction of the electric field. Such materials are called *isotropic*. If the conductivity is different in different directions, the material is called *anisotropic*, a very rare condition.

We will consider only uniform, isotropic materials for which $\sigma = $ constant. As in single variable calculus, σ can be factored out of the derivative to give

$$\nabla \cdot (\sigma \mathbf{E}) = \sigma \nabla \cdot \mathbf{E} = 0. \tag{1.118}$$

Note the vector identity in Appendix A for the case where σ is not constant.
Substituting Eq. (1.114) into Eq. (1.117) and dividing both sides by σ leads to

$$\nabla \cdot \mathbf{E} = \nabla \cdot (-\nabla V) = -\nabla \cdot \nabla V = 0. \tag{1.119}$$

We now have a single equation in only one variable that expresses the behavior of the voltage throughout the resistor. The form of this equation depends upon the coordinate system used. In Cartesian coordinates, Eq. (1.119) becomes

$$\nabla \cdot \nabla V = \left(\mathbf{a}_X \frac{\partial}{\partial x} + \mathbf{a}_Y \frac{\partial}{\partial y} + \mathbf{a}_X \frac{\partial}{\partial z} \right) \cdot \left(\mathbf{a}_X \frac{\partial V}{\partial x} + \mathbf{a}_Y \frac{\partial V}{\partial y} + \mathbf{a}_X \frac{\partial V}{\partial z} \right)$$
$$= \frac{\partial^2 V}{\partial x^2} + \frac{\partial^2 V}{\partial y^2} + \frac{\partial^2 V}{\partial z^2} = 0, \tag{1.120}$$

a PDE for the voltage within the resistor. This form has been widely studied since it appears in many areas of physics. Mathematicians have given a special name and symbol to this form. It is called the *Laplacian* in honor of Pierre Simon de Laplace, an 18th century pioneer in mathematics and physics. It is denoted by the symbol ∇^2. Since the gradient of V, ∇V, is a vector, we can interpret the Laplacian as the divergence of the gradient; the result is a scalar. In Cartesian coordinates, the Laplacian is defined as

$$\nabla^2 V = \frac{\partial^2 V}{\partial x^2} + \frac{\partial^2 V}{\partial y^2} + \frac{\partial^2 V}{\partial z^2}. \tag{1.121}$$

The forms of the Laplacian in cylindrical and spherical coordinate systems are given in Appendix C. Most of our work will be confined to Cartesian coordinates.

Equation (1.120) is written in terms of the Laplacian as

$$\nabla^2 V = \frac{\partial^2 V}{\partial x^2} + \frac{\partial^2 V}{\partial y^2} + \frac{\partial^2 V}{\partial z^2} = 0. \tag{1.122}$$

This is known as *Laplace's equation*. Since the RHS is zero, it is a homogeneous PDE—a mathematical indication that there are no sources of flux within the resistive material. This is not too surprising since we already know that $\nabla \cdot \mathbf{J} = 0$ that implies there can be no sources.

Finally, an interesting situation occurs when charge can accumulate within the conductive region—a condition which cannot occur within resistors. For this case, Eq. (1.114) and (1.116) apply as before, but Eq. (1.85) must be used due to the accumulation of charge as

$$\nabla \cdot \mathbf{J} = -\frac{\partial \rho_V}{\partial t}. \tag{1.123}$$

Combining these equations as before, we obtain

$$\nabla^2 V = \frac{1}{\sigma} \left(\frac{\partial \rho_V}{\partial t} \right). \tag{1.124}$$

This equation, known as *Poisson's equation*, is inhomogeneous since the RHS is not zero indicating the presence of sources of flux. The charge density frequently depends upon the voltage, further complicating the solution. Fortunately, Poisson's equation doesn't occur very frequently in passive circuit elements so we rarely have to solve it.

1.29   ANALYTIC SOLUTION METHODS

The solution of Laplace's equation by analytic methods is practical only in Cartesian, cylindrical, and spherical coordinate systems. Though very few problems actually fit these coordinate systems, their solutions provide a model or guide as to the form of the voltage solution in similar configurations. To gain some insight into these methods, we will obtain solutions for a couple of geometries.

Let's apply Eq. (1.122) to the axial resistor we have considered earlier, shown again in Fig. 1.24. With the electrodes separated by a distance L in the z-direction, a voltage difference V_S is established in this direction. However, there is no voltage drop in the x- and y-directions since the PEC electrodes with zero tangential electric field lie in these directions. Consequently, $\partial V/\partial x = \partial V/\partial y = 0$ and Eq. (1.122) reduces to

$$\frac{\partial^2 V}{\partial z^2} = 0. \tag{1.125}$$

Since the voltage varies only with respect to z, the partial derivatives can be replaced by ordinary derivatives. Direct integration enables us to obtain the analytic form of V as

$$V(z) = Az + B. \tag{1.126}$$

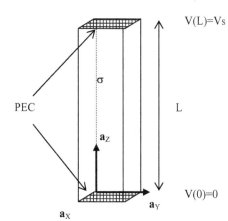

FIGURE 1.25: Axial resistor.

The two arbitrary constants are determined by the boundary conditions, i.e., the voltage at the two boundaries of the resistor to which the electrodes are attached. From Fig. 1.25, we see that $V(L) = V_S$ and $V(0) = 0V$. Applying these values to Eq. (1.126), we solve for $A = V_S/L$ and $B = 0$ which leads to

$$V(z) = \frac{V_S z}{L}. \tag{1.127}$$

Not surprisingly, the voltage varies linearly with z resulting in an accompanying electric field intensity of

$$\mathbf{E} = -\nabla V = -\frac{V_S \mathbf{a}_Z}{L} \tag{1.128}$$

and a current density of

$$\mathbf{J} = \sigma \mathbf{E} = -\frac{V_S \sigma \mathbf{a}_Z}{L}. \tag{1.129}$$

Integrating the current density over any cross-sectional surface of the resistor, say the electrode at $z = 0$, we obtain the current flowing through the resistor as

$$I_R = \iint_A \mathbf{J} \cdot \mathbf{ds} = -\frac{V_S \sigma}{L} \iint_A \mathbf{a}_Z \cdot (-\mathbf{a}_Z) dx dy$$

$$= \frac{V_S \sigma A}{L}. \qquad (1.130)$$

Finally, the resistance is given as

$$R = \frac{V_R}{I_R} = \frac{V_S}{I_R} = \frac{V_S}{V_S \sigma A / L} = \frac{L}{\sigma A}. \qquad (1.131)$$

This is exciting! We have confirmed the formula used in physics, but in obtaining these results we have developed a consistent model for the internal behavior of resistors. We have defined microscopic electromagnetic parameters that lead to observed behavior of resistors.

The mathematical features of the voltage reveal additional information. The first and second derivatives of the voltage are given by

$$\frac{dV}{dz} = \frac{V_S}{L} \qquad (1.132)$$

and

$$\frac{d^2 V}{dz^2} = 0, \qquad (1.133)$$

respectively. The constant value of the first derivative of voltage with respect to z implies that the voltage varies linearly within the resistor from a minimum value at $z = 0$ to a maximum value at $z = L$. As you recall from calculus, there is no maximum or minimum at points where the second derivative vanishes. Equation (1.133) implies that *there is no maximum or minimum of the voltage within the resistor*; these occur at the electrodes on each end of the resistor. Though the solutions are more complicated for other configurations, they all show the same behavior— there are no maxima or minima within the conductive material of the resistor. These occur only at the electrodes.

Variations of the voltage with respect to a single variable can occur in cylindrical and spherical coordinate systems too. Direct integration is an appropriate solution method for these problems as well. Whenever the voltage depends upon more than one spatial coordinate, a Laplace's equation becomes a PDE. The most common analytic solution method is known as the separation of variables. These techniques are covered in typical PDE course, but they are beyond the scope of this textbook. These methods are reserved for advanced electromagnetics courses.

Example 1.29-1.   Calculate the resistance of a coaxial resistor of length L which is filled with conductive material σ between PEC cylindrical electrodes of inner radius of $\rho = a$ and outer radius of $\rho = b$. In order to calculate resistance, we assume a voltage drop of V_O and calculate the resultant current flow. To accomplish this we need to find $\mathbf{J}$ which we can calculate from $\mathbf{E} = -\nabla V$. So we must solve for $V(\rho, \phi, z)$. There are no axial variations within the resistor. Consequently, we will assume that within this region the fields are identical to those of an infinitely long resistor, i.e., $\partial V / \partial z = 0$. Moreover, the structure has angular symmetry so the fields are expected to be the same for all values of ϕ so that $\partial V / \partial \phi = 0$. Therefore, Laplace's equation becomes $\nabla^2 V = \frac{1}{\rho} \frac{\partial}{\partial \rho} \left(\rho \frac{\partial V}{\partial \rho} \right) = 0$. Integrating with respect to ρ, we obtain $\rho \frac{\partial V}{\partial \rho} = A$ which can be rewritten as $\frac{\partial V}{\partial \rho} = \frac{A}{\rho}$. A second integration gives $V = A \ln(\rho) + B$. Let's assume that the outer electrode is grounded, $V(b) = 0$, and that the inner has a voltage V_O, $V(a) = V_O$. Applying these boundary conditions, we calculate $A = V_O / \ln(a/b)$ and $B = -V_o \ln(b) \ln(a/b)$ so that $V(\rho) = \frac{V_O}{\ln(a/b)} (\ln(\rho) - \ln(b)) = V_O \frac{\ln(\rho/b)}{\ln(a/b)}$. From this expression, we find $\mathbf{E} = -\nabla V$ as

$$\mathbf{E} = -\nabla V = -\frac{\partial V}{\partial \rho} \mathbf{a}_\rho = -\frac{V_O}{\ln(a/b)\rho} \mathbf{a}_\rho = \frac{V_O}{\ln(b/a)\rho} \mathbf{a}_\rho.$$

As expected, $\mathbf{E}$ points away from the more positive inner electrode in the outward radial direction. The current flux points in the same direction since $\mathbf{J} = \sigma \mathbf{E}$. The resistor current is calculated by integrating over all the $\mathbf{J}$ within the resistor. Any cylindrical surface extending from one end to the other will suffice as

$$I_R = \iint_S \mathbf{J} \cdot \mathbf{ds} = \int_{z=0}^{L} \int_{\phi=0}^{2\pi} \frac{\sigma V_O}{\ln(b/a)\rho} \mathbf{a}_\rho \cdot \mathbf{a}_\rho \rho \, d\phi \, dz$$

$$= \frac{\sigma V_O L}{\ln(b/a)} \int_{\phi=0}^{2\pi} d\phi = \frac{2\pi \sigma V_O L}{\ln(b/a)}.$$

At last, we can calculate the resistance of a cylindrical resistor with electrodes on inner and outer cylindrical surfaces as

$$R = \frac{V_R}{I_R} = \frac{V_O}{\frac{2\pi \sigma V_O L}{\ln(b/a)}} = \frac{\ln(b/a)}{2\pi \sigma L}.$$

Due to the cylindrical geometry, the "length" between the two electrodes is $\ln(b/a)$. This procedure can be applied to other geometries as well.

1.30 NUMERIC METHODS

In general, analytic solution methods are practical in Cartesian, cylindrical, and spherical coordinate systems only. Once an analytic solution is obtained, the voltage can be evaluated exactly at any point within the solution region. More sophisticated analytic techniques required for other configurations are beyond the scope of this text. Alternatively, numeric methods can be applied to virtually any geometry, though they provide approximate solutions that are valid at only a finite number of points. Typically, the voltages are calculated only at the nodes (or intersection points) of a rectangular grid which subdivides the solution region. Linear interpolation of node voltages can be used to calculate the voltage at other points. Greater solution accuracy is achieved with finer grids and more nodes, but at the expense of increased computational effort.

Powerful numeric electromagnetic solvers have replaced the analytic solutions methods. Though much more sophisticated than the ele mentary methods that follow, they have their basis in the same fundamental laws that we have developed.

The simplest numeric methods are based upon incremental approximations of Laplace's equation. A typical grid structure is shown in Fig. 1.26 with grid spacings of Δx, Δy, and Δz in the x-, y-, and z-directions, respectively. The central node is denoted as O; the surrounding nodes—up, down, front, back, left, and right— are denoted as U, D, F, B, L, and R, respectively. The first derivative of voltage with respect to x is approximated at point A as

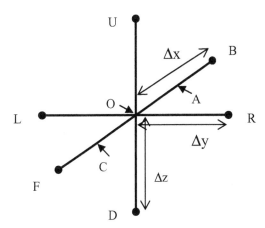

FIGURE 1.26: Basic cell for numeric methods.

$$\left.\frac{\partial V}{\partial x}\right|_A \approx \frac{V_O - V_B}{\Delta x} \tag{1.134}$$

and at point C as

$$\left.\frac{\partial V}{\partial x}\right|_C \approx \frac{V_F - V_O}{\Delta x} \tag{1.135}$$

where V_i is the voltage at the ith node. This is known as the central difference form since the increment Δx is centered on the point of interest. No doubt you are more familiar with the traditional forward difference form of differential calculus in which the increment precedes or is forward from the point of interest. The central difference form is preferred for numerical computations, but is theoretically the same. As the increment decreases, the results of the two forms become indistinguishable.

With the same technique, the second derivative of voltage at point O is approximated by

$$
\begin{aligned}
\frac{\partial^2 V}{\partial x^2}\bigg|_O &\approx \frac{\frac{\partial V}{\partial x}\big|_C - \frac{\partial V}{\partial x}\big|_A}{\Delta x} \\
&= \frac{\frac{V_F - V_O}{\Delta x} - \frac{V_O - V_B}{\Delta x}}{\Delta x} \\
&= \frac{V_F + V_B - 2V_O}{(\Delta x)^2}.
\end{aligned}
\tag{1.136}
$$

The second derivative of the voltage at point O is expressed in terms of the voltages to the left, to the right, and at the point O and of the increment between nodes. Applying this procedure to y- and z-variations, we obtain

$$
\frac{\partial^2 V}{\partial y^2}\bigg|_O \approx \frac{V_L + V_R - 2V_O}{(\Delta y)^2}
\tag{1.137}
$$

and

$$
\frac{\partial^2 V}{\partial z^2}\bigg|_O \approx \frac{V_U + V_D - 2V_O}{(\Delta z)^2}.
\tag{1.138}
$$

Equations (1.136), (1.137), and (1.138) combine to give the incremental form of the Laplacian. For convenience let $\Delta x = \Delta y = \Delta z = \Delta$ so that the Laplacian at point O is particularly simple as

$$
\nabla^2 V \approx \frac{V_F + V_B + V_R + V_L + V_U + V_D - 6V_O}{\Delta^2} = 0.
\tag{1.139}
$$

For nonzero Δ, the voltage at node O can be obtained by solving Eq. (1.140) as

$$
V_O = \frac{V_F + V_B + V_R + V_L + V_U + V_D}{6}.
\tag{1.141}
$$

This is an extremely useful equation. It describes a fundamental property of the voltage within a resistor: *the voltage at any point is the average of the voltage of surrounding points.* This reaffirms our earlier finding that there can be no maximum within the solution region. This is valid within any conducting material in which no charge accumulates. In addition, it is the basis for several methods of numeric calculations of the voltage. This is the numeric form of Laplace's equation!!

As a general rule, the smaller the increment Δ, the more accurately the incremental Laplacian approximates the differential form. But to keep the number of computations reasonable, Δ

should not be too small. A general guideline similar to that for curvilinear squares is applicable—at least three or four nodes in the direction of the smallest dimension of the solution region should give 5–10% accuracy in the node voltages.

The incremental form of Laplace's equation can be expressed in general form for an N-dimensional problem as

$$V_O = \frac{\sum_{i=1}^{N} V_i}{2N} \tag{1.142}$$

where V_i is the voltage at the surrounding nodes. For $N = 1$, there are only two nearby nodes, for $N = 2$ there are four, and for $N = 3$ there are six.

Some problems are more naturally suited to other coordinate systems; incremental forms can be obtained in cylindrical and spherical coordinate systems as well with somewhat more complicated incremental forms. However, the simplicity of calculations in Cartesian coordinates makes them most attractive for our use in this text.

Several methods of calculating voltages based upon the incremental Laplacian are described in the following sections. The two-dimensional problem of Fig. 1.27 will be repeated to illustrate these methods. The solution region is subdivided by an equispaced grid. Division of the shortest dimension of the structure into four cells produces a 3×3 grid with nine nodes which is superimposed on the structure.

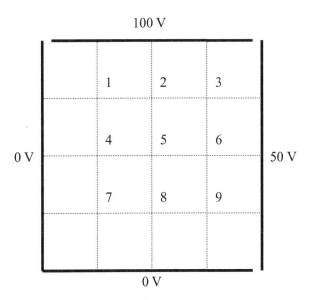

FIGURE 1.27: Numeric calculation of node voltages.

1.31 LINEAR EQUATIONS

The incremental Laplacian of Eq. (1.142) is valid at all points within the solution region of the resistor. Therefore, if we impose it simultaneously at all nine unknown nodes of the resistor, we will obtain nine equations in nine unknowns. This linear set of equations can be solved readily by numeric means to obtain the node voltages. Imposing Laplace's equation at node 1, we obtain

$$V_1 = \frac{100 + 0 + V_2 + V_4}{4}, \tag{1.143}$$

which can be rewritten as

$$4V_1 - V_2 - V_4 = 100. \tag{1.144}$$

At node 2, we obtain

$$V_2 = \frac{100 + V_5 + V_1 + V_3}{4}, \tag{1.145}$$

which can be rewritten as

$$-V_1 + 4V_2 - V_3 - V_5 = 100. \tag{1.146}$$

For node 9, we obtain

$$-V_6 - V_8 + 4V_9 = 50. \tag{1.147}$$

Careful observations show that a pattern for the equation at the Nth node includes the following terms on the RHS: $4V_N$ minus the sum of the surrounding unknown nodal voltages. The LHS contains the sum of the boundary voltages adjacent to the Nth node. With this pattern, we can quickly write each of the nodal equations. A further simplification is to combine the nine nodal equations into matrix form as

$$[C][NV] = [BV] \tag{1.148}$$

where $[NV]$ is a 9×1 column matrix with each entry representing one of the nine nodal voltages, $[C]$ is the configuration matrix, a 9×9 square matrix with elements determined by the problem geometry, and $[BV]$ is a 9×1 column matrix determined by the boundary voltages. For the configuration shown in Fig. 1.27, these matrices are given by

$$[NV] = \begin{bmatrix} V_1 \\ V_2 \\ V_3 \\ V_4 \\ V_5 \\ V_6 \\ V_7 \\ V_8 \\ V_9 \end{bmatrix}, \quad [BV] = \begin{bmatrix} 100 \\ 100 \\ 150 \\ 0 \\ 0 \\ 50 \\ 0 \\ 0 \\ 50 \end{bmatrix}, \quad \text{and}$$

$$[C] = \begin{bmatrix} 4 & -1 & 0 & -1 & 0 & 0 & 0 & 0 & 0 \\ -1 & 4 & -1 & 0 & -1 & 0 & 0 & 0 & 0 \\ 0 & -1 & 4 & 0 & 0 & -1 & 0 & 0 & 0 \\ -1 & 0 & 0 & 4 & -1 & 0 & 0 & 0 & 0 \\ 0 & -1 & 0 & -1 & 4 & -1 & 0 & -1 & 0 \\ 0 & 0 & -1 & 0 & -1 & 4 & 0 & 0 & -1 \\ 0 & 0 & 0 & -1 & 0 & 0 & 4 & -1 & 0 \\ 0 & 0 & 0 & 0 & -1 & 0 & -1 & 4 & -1 \\ 0 & 0 & 0 & 0 & 0 & -1 & 0 & -1 & 4 \end{bmatrix}, \qquad (1.150)$$

respectively. Fundamental matrix operations lead to the solution for the node voltages as

$$[NV] = [C]^{-1}[BV]. \qquad (1.151)$$

The configuration matrix $[C]$ has several interesting properties governed by the nature of the incremental Laplacian. The matrix is symmetric about the upper-left to lower-right diagonal, i.e., $C_{ij} = C_{ji}$. In the language of matrix theory, $[C] = [C]^T$, the matrix equals its own transpose. This is because the effects of the potentials at the ith and jth nodes on each other are the same. The matrix shows diagonal dominance, i.e., the diagonal term is greater than or equal to the sum of all the other terms in that row or column. The matrix is sparse, i.e., there are relatively few nonzero elements. Since there are at most five nonzero elements in each row and column due the node and its four surrounding nodes, the sparsity of $[C]$ increases as the number of nodes increases.

$[C]^{-1}$ is rarely calculated directly since this requires excessive memory and computation time. Gaussian elimination or similar techniques are popular for matrices up to several dozen nodes. However, sparse matrix methods are required for problems with hundreds or more nodes. Details of numeric analysis are beyond the scope of the text; but they can be found in the references.

For the problem of Fig. 1.27, the calculation of node voltages gives

$$[NV] = \begin{bmatrix} 45.5 \\ 61.4 \\ 64.0 \\ 20.4 \\ 36.3 \\ 44.7 \\ 9.8 \\ 18.6 \\ 28.3 \end{bmatrix}. \tag{1.152}$$

An exact solution obtained by more advanced analytic methods gives the result of

$$[NV]_{\text{EXACT}} = \begin{bmatrix} 46.6 \\ 63.2 \\ 64.8 \\ 23.0 \\ 37.5 \\ 45.2 \\ 10.2 \\ 18.6 \\ 28.4 \end{bmatrix}. \tag{1.153}$$

A comparison of Eqs. (1.152) and (1.153) shows that the nodal voltages are within 5% of the exact solution, not a bad approximation.

The voltage at a non-nodal location within the solution region can be approximated most simply by using linear interpolation between nearby nodes. Consider the four nodes shown in Fig. 1.28 and assume that the voltage varies linearly in both the x- and y-directions, i.e.,

$$V_P = (A\Delta x + B)(C\Delta y + D) \tag{1.154}$$

Evaluation of V_P at each of the known nodal voltages provides four equations in four unknowns which is solved and the results substituted into Eq. (1.154) to give

$$V_P = V_A\left(1 - \frac{\Delta x}{\Delta} - \frac{\Delta y}{\Delta} + \frac{\Delta x \Delta y}{\Delta^2}\right) + V_D\frac{\Delta x \Delta y}{\Delta^2}$$

$$+ V_B\frac{\Delta y}{\Delta}\left(1 - \frac{\Delta x}{\Delta}\right) + V_C\frac{\Delta x}{\Delta}\left(1 - \frac{\Delta y}{\Delta}\right). \tag{1.155}$$

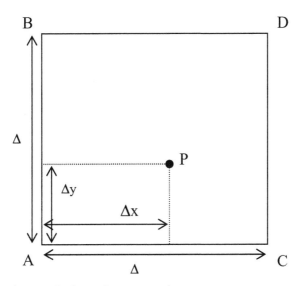

FIGURE 1.28:   Interpolation of voltages between nodes.

This result is sufficiently general to approximate the voltage at any point in the solution region. Note that it assumes a linear interpolation between the approximate nodal values. Nevertheless, this method gives adequate results for our studies. For increased accuracy, the size of the increment Δ can be reduced. We will reserve more sophisticated approximations for advanced courses.

Example 1.31-1.   Estimate the voltage at a point centrally located between nodes 1, 2, 4, and 5. For this point $\Delta x/\Delta = \Delta y/\Delta = 0.5$ and the voltages are $V_A = 20.4$, $V_B = 45.5$, $V_C = 36.3$, and $V_D = 61.4$. Application of Eq. (1.155) to these data gives

$$V_P = 20.4(1 - 0.5 - 0.5 + 0.25) + 61.4(0.25)$$
$$+ 45.5(0.5)(1 - 0.5) + 36.3(0.5)(1 - 0.5) = 40.9 \text{ V.}$$

The solution of Eq. (1.152) can be used to sketch equipotentials throughout the region as shown in Fig. 1.29. Moreover, we can find **E** by using its basic definition as a magnitude of $|\Delta V/\Delta l|$ pointing in the direction from more positive to less positive values of V. These results are sketched in Fig. 1.29 also.

1.32   ITERATIVE TECHNIQUES

Another numeric method iteratively applies the incremental Laplacian to each node; the result converges to the same solution obtained by the linear equation method of the previous section.

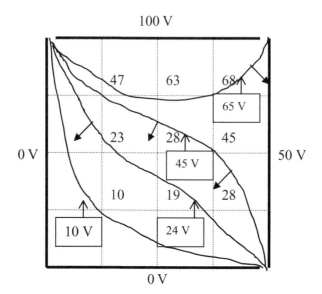

FIGURE 1.29: Voltage and electric field by linear equations and interpolation.

The process begins with a set of "guessed" node voltages. The systematic, sequential application of Eq. (1.142) to each node brings the node voltages closer to the correct solution. The sequence is repeated again using the improved values with even better results. This process is iterated until there are no further changes of the node voltages within the desired accuracy. This very simple solution method is possible due to the diagonally dominant nature of the configuration matrix $[C]$ that ensures convergence of the process to the correct solution. This method is also known as the *relaxation method*; the node voltages "relax" to the correct value.

The number of iterations can be reduced by making good guesses for the original node voltages. However, even the assumption that all the nodes are at zero volts still results in the correct solution. The poorer the guess, the more iterations required. When computations are by hand, it is especially important that the guessed node voltages be as accurate as possible to reduce the number of iterations. When the method is implemented via a computer, it may be easier to increase the number of iterations rather than spend effort making good guesses.

The iterative method is especially effective for closed regions where the voltages on the entire boundary surrounding the solution region are known. In general, open problems that extend to infinity present a significant problem as an infinite number of nodes are required. But for the cases where the conductive material can be treated as an ideal flux guide, iterative solutions work quite well. By imposing the special boundary condition that no current leaves the edges of the resistor, we do not need to include any nodes outside the conductive material. Details on special techniques for more general open problems and other enhancements of the method are in the references.

The details of this method will be clearer by two examples—one with hand computations and the other via computer.

Example 1.32-1.   Calculate the node voltages of the closed region of Fig. 1.27 by hand calculations. Since we want to minimize calculations by hand, we must make good initial guesses of the node voltages. The four corner points are equally distant from node 5 and provide the basis for estimating its value as $V_5 = (50 + 75 + 25 + 0)/4 \approx 38$ V. Note that at each corner the voltage is taken as the average of the two adjacent electrode voltages. Since we don't expect accuracy greater than 1%, we will round off all node voltages to integer values. The electrodes and the estimated V_5 enable us to estimate V_1, V_3, V_7, and V_9 as $V_1 = (50 + 100 + 38 + 0)/4 = 47$, $V_3 = 66$, $V_7 = 10$, and $V_9 = 28$. These voltages are used in turn to estimate the remaining node voltages as $V_2 = 63$, $V_4 = 24$, $V_6 = 46$, and $V_8 = 19$. These estimated values are shown in Fig. 1.30. The size and orientation of the grid used in making the estimates varies from node to node. Other grids could be chosen with somewhat different results. But, since this process is just obtaining reasonable estimates, these is no "correct" result, though generally, the smaller the grid spacing, the more accurate the estimate. Regardless of the estimated values, the final results converge to the correct solution obtained by the linear equations method. A comparison of these estimates with the exact solution of Eq. (1.140) reveals that they are remarkably accurate ($\approx 2\%$) without any further calculations. This accuracy is a reflection of the power of the concept that nodal voltages are the average of adjacent nodal voltages. The sequential application of Eq. (1.142) at each node should follow a pattern. For convenience, let's just follow the node numbers in sequence. Each calculation will use the most recent voltages for each of the nodes. A good system for tracking the process is to cross out the previous value at a node and enter the new one below it as illustrated in Fig. 1.31.

Since we made some very good estimates, the sequence of iterations produced only minor changes to the estimated values. This method is rather immune to errors since no matter what values of node voltages are used in a step, the results will approach the correct solution. Errors merely slow the process and necessitate more iterations.

Example 1.32-2.   Calculate the node voltages of the closed region of Fig. 1.27 with the aid of a computer. The repetitive application of Eq. (1.142) at each node required by this method is easily implemented with a spreadsheet. Each of the nodes at the center of a cell in Fig. 1.27 corresponds to a cell in the Excel spreadsheet in Fig. 1.32.

The cells with values of 0, 50, or 100 are the electrode with fixed voltages. The voltages in the interior cells are calculated by Eq. (1.142). The voltage of the highlighted cell in

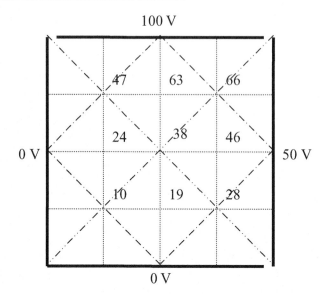

FIGURE 1.30: Estimated node voltages for the iteration method.

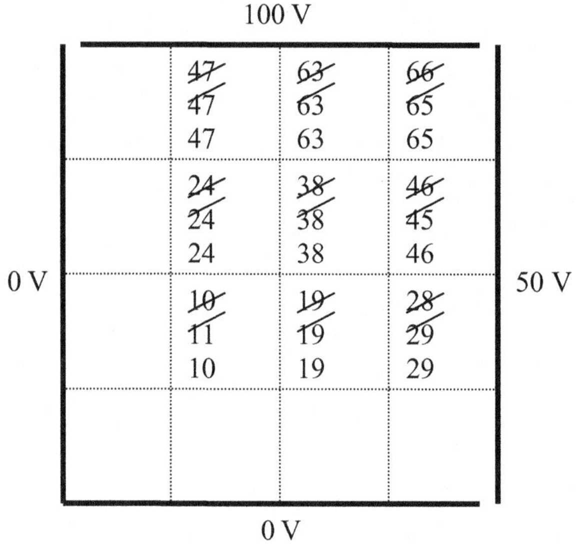

FIGURE 1.31: Hand calculation procedures of the iteration method.

Fig. 1.32, cell C5, is calculated from the voltages of the surrounding cells by the cell formula of "=(B5+C4+D5+C6)/4." This formula is replicated easily in the other interior cells using the COPY command. The calculations are iterated until no integer changes of the voltages occur. Note that Excel will notify you about "circular reference" where cell AA5 refers to cell AA4

B	C	D	E	F
2 Voltage Calculation by Iteration Method				
3				
4	100	100	100	
5	0	46	62	64	50
6	0	24	38	45	50
7	0	11	19	29	50
8		0	0	0	

FIGURE 1.32: Spreadsheet calculations using the iteration method.

which in turn refers to cell AA5. The Tools/Options menu enables selection of iteration details. The results, displayed within the each cell, agree very well with the exact results of Eq. (1.153).

1.33 RESISTANCE CALCULATIONS

The calculation of node voltages poses special problems at the open edges of a resistor. However, when the conductive material acts as an ideal flux guide, modification of the process is quite simple. In this case no current can leave the resistor from the edges. All of the current must be tangential to the boundary and the normal component must be zero, i.e., $J_N = 0$. Putting this in terms of voltage,

$$J_N = \sigma E_N = -\sigma \frac{\partial V}{\partial n} = 0, \tag{1.156}$$

we find that the derivative of the voltage in the direction of the normal to the edge must vanish at each edge. To implement this condition, we define a fictitious, auxiliary node, V_{OUT}, adjacent to the boundary as shown in Fig. 1.33.

Of course, Eq. (1.142) must hold at point O. In addition, the normal derivative is approximated by

$$\left.\frac{\partial V}{\partial n}\right|_E \approx \frac{V_{OUT} - V_{IN}}{2\Delta} = 0, \tag{1.157}$$

which leads to

$$V_{OUT} = V_{IN}. \tag{1.158}$$

By forcing the node voltage of the fictitious node outside of the resistor to equal the voltage at the interior node, we cause the normal derivative to vanish at the edge of the resistor,

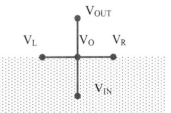

FIGURE 1.33: Basic cells at the edges of conductive region.

node E. Substitution of $V_{\text{OUT}} = V_{\text{IN}}$ into Eq. (1.142) gives

$$V_E = \frac{V_L + V_R + V_{\text{IN}} + V_{\text{OUT}}}{4}$$
$$= \frac{V_L + V_R + 2V_{\text{IN}}}{4} \qquad (1.159)$$

which is imposed at all edges of the resistor where there are only tangential currents.

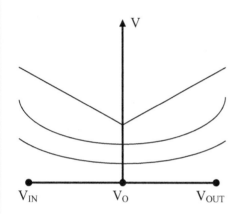

FIGURE 1.34: Node voltages at the edge of a resistor.

Since the voltage of the fictitious node is equal to the voltage of the symmetrically located internal node, $V_{\text{OUT}} = V_{\text{IN}}$, the voltages surrounding the edge node V_O are symmetric. If the voltage on either side of a point is symmetric, then the voltage at that point is either a maximum or minimum. This implies a zero derivative of V_O as required. This is illustrated in Fig. 1.34.

The behavior of the fields at interior and exterior corners requires more detailed analyses of this sort. Instead, let us use an alternate method of calculating the nodal relationships which is valid for all edges between different materials. Using the continuity of current as the guiding principle, we can obtain results that can be used for materials with conductivity that varies from point to point, i.e., inhomogeneous materials.

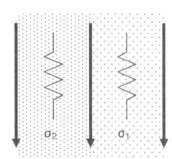

FIGURE 1.35: Interface parallel with current flux lines.

When a material interface is parallel to flux lines, no flux will cross from one material into another; the current will be completely tangential to the interface. Furthermore, the equality of the tangential electric fields ensures that the voltage drops across the two materials are the same. The two conductive materials act as two separate resistors with equal voltage drops that share common electrodes. But, this is just two resistors in parallel! Indeed, we can calculate each of the resistances independently and then find the total resistance of the two parallel resistors, see Fig. 1.35. When an interface is parallel to an equipotential, the flux lines are perpendicular to the boundary and continuous from one material into the other. Hence, both regions conduct the same current. A PEC electrode could be inserted upon the equipotential interface between the two materials without altering the flux lines. This means that the two conductive materials

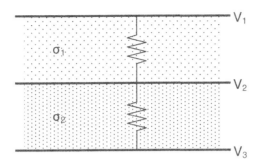

FIGURE 1.36: Interface parallel with equipotentials.

act as two resistors in series. The separate voltage drops and resistances can be calculated. The total resistance is calculated by the series resistance rule, see Fig. 1.36.

A more challenging case is when the interface is arbitrarily positioned; it is parallel to neither flux lines nor equipotentials. Analytically, this is beyond the scope of this text. But it is quite possible for us to handle this situation numerically. A first step is to discretize the boundary into a series of straight-line segments which are coincident with the center lines of basic computation cells. In the most general case, a cell has different material in each quadrant, see Fig. 1.37.

Laplace's equation, Eq. (1.142), is based upon spatial derivatives. Unfortunately, they are not finite at material interfaces and this is not a valid solution method. An alternate relationship

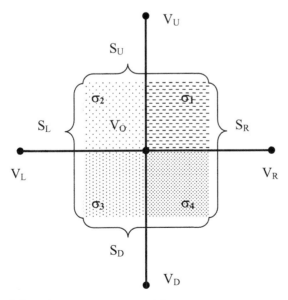

FIGURE 1.37: Basic cell for inhomogeneous materials.

between the node voltages is provided by the continuity of current, Eq. (1.48), which is valid everywhere. The integral is expressed in terms of the unit cell as

$$
I_{\text{OUT}} = \oiint_S \mathbf{J} \cdot \mathbf{ds} = \sum_{i=1}^{N} \iint_{S_i} \mathbf{J} \cdot \mathbf{ds} = \sum_{i=1}^{N} \iint_{S_i} \sigma \mathbf{E} \cdot \mathbf{ds}
$$

$$
= -\sum_{i=1}^{N} \iint_{S_i} \sigma \nabla V \cdot \mathbf{ds} = -\sum_{i=1}^{N} \iint_{S_i} \sigma \nabla V \cdot \mathbf{a}_N da
$$

$$
= -\sum_{i=1}^{N} \iint_{S_i} \sigma \frac{\partial V}{\partial n} da = 0 \tag{1.160}
$$

where S_i is the ith surface of the computation cell. Each of the surface integrals represents the current flux that crosses the ith surface. Expressed in incremental form, the ith surface integral is approximated as

$$
\iint_{S_i} \sigma \frac{\partial V}{\partial n} da = \iint_{S_{i1}} \sigma_{\text{ia}} \frac{\partial V}{\partial n} da + \iint_{S_{i2}} \sigma_{\text{ib}} \frac{\partial V}{\partial n} da
$$

$$
\approx \sigma_{\text{ia}} \frac{V_i - V_O}{\Delta} \frac{\Delta t}{2} + \sigma_{\text{ib}} \frac{V_i - V_O}{\Delta} \frac{\Delta t}{2}
$$

$$
= \left[\frac{\sigma_{\text{ia}} + \sigma_{\text{ib}}}{2} \right] t(V_i - V_O)
$$

$$
= \sigma_{\text{iAVE}} t(V_i - V_O) \tag{1.161}
$$

where σ_{iAVE} is the average conductivity at the ith surface of the computation cell. The cell has equal height and width of Δ and a thickness into the page of t. The electric field intensity is approximated as $(V_i - V_O)/\Delta$. Current density through the ith surface is just the electric field intensity times the conductivity, $\sigma_{\text{ia}}(V_i - V_O)/\Delta$ and $\sigma_{\text{ib}}(V_i - V_O)/\Delta$ for materials a and b, respectively. Due to the small size of the cell, the current density is constant over the ith surface and the integral is approximated as the current density times the area of the ith surface $\Delta I_i = [(\sigma_{\text{ia}} + \sigma_{\text{ib}})(V_i - V_O)/\Delta](\Delta/2)t = \sigma_{\text{iAVE}}(V_i - V_O)t$. Since current flows into the cell when $V_i > V_O$, the minus sign is required and the continuity equation becomes

$$
I_{\text{OUT}} = 0 = -\sum_{i=1}^{N} \iint_{S_i} \sigma \frac{\partial V}{\partial n} da
$$

$$
\approx -t \sum_{i=1}^{N} \sigma_{\text{iAVE}}(V_i - V_O). \tag{1.162}
$$

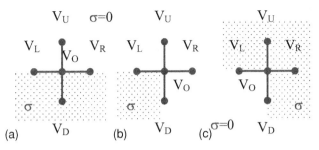

FIGURE 1.38: Boundary points for conductive regions of resistors: (a) edges, (b) exterior corners, and (c) interior corners.

This equation can be solved for V_O in terms of the adjacent voltages V_i as

$$V_O = \frac{\sigma_{UAVE} V_U + \sigma_{LAVE} V_L + \sigma_{DAVE} V_D + \sigma_{RAVE} V_R}{(\sigma_1 + \sigma_2 + \sigma_3 + \sigma_4)}. \tag{1.163}$$

Wow, that was a lot of work! But we are now prepared to obtain numeric solutions for regions with inhomogeneous conductivity. Situations with continuous variations in conductivity can be solved with Eq. (1.163) also. In these cases, the value of σ at the center of each surface replaces $\sigma_{iAVE} = (\sigma_{ia} + \sigma_{ib})/2$. Note that when $\sigma_1 = \sigma_2 = \sigma_3 = \sigma_4 = \sigma$, Eq. (1.163) reduces to Eq. (1.142) as it should. Our favorite numeric methods—linear equations, iteration, and, yet to come, circuit simulation—can use Eq. (1.163). Using Eq. (1.163), we can also obtain the cell formulas for the several types of resistor boundaries.

With the edge oriented as in Fig. 1.38(a) and with the resistor acting as an ideal flux guide, i.e., $\sigma_1 = \sigma_2 = 0$ and $\sigma_3 = \sigma_4 = \sigma$, Eq. (1.163) becomes

$$V_E = \frac{V_L + 2V_D + V_R}{4} \tag{1.164}$$

in agreement with Eq. (1.159) obtained earlier by an other method. For an exterior corner as shown in Fig. 1.38(b), $\sigma_1 = \sigma_2 = \sigma_4 = 0$ and $\sigma_3 = \sigma$, so the cell formula becomes

$$V_{EC} = \frac{V_L + V_D}{2}. \tag{1.165}$$

For an interior corner as shown in Fig. 1.38(c), $\sigma_1 = \sigma_2 = \sigma_4 = \sigma$ and $\sigma_3 = 0$, so the cell formula becomes

$$V_{IC} = \frac{2(V_U + V_R) + V_D + V_L}{6}. \tag{1.166}$$

The resistor current, I_R, must be determined in order to calculate the resistance value. Again, we can use the principles we have learned so far to perform this calculation. The current

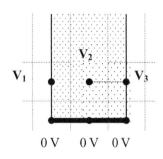

FIGURE 1.39: Calculation of the resistor current.

flowing through any cross section of the resistor must be calculated from the node voltages. For convenience, consider the shaded cells adjacent to the zero-volt electrode in Fig. 1.39. The current flowing into the electrode from the ith interior cell can be calculated as

$$\Delta I_i = \iint_{\Delta A_i} \mathbf{J} \cdot \mathbf{ds} \approx \sigma \left(\frac{V_i - V_{\text{ELECT}}}{\Delta} \right) \Delta t = \sigma t V_i \quad (1.167)$$

where the $V_{\text{ELECT}} = 0$ is the voltage of the electrode, Δ is the width of each cell, t is the thickness of the resistor, and Δt is the area of the ith surface of each cell. In the two edge cells, $i = 1$ and $i = N$, current only flows through the half of the cell which is within the conductive material. Therefore, the area is only half of the cell width, $A_1 = A_N = \Delta t/2$.

The resistor's current is the sum of the currents flowing from each cell into the adjacent zero-volt electrode. The resistor current is expressed as

$$I_R \approx \sum_{i=1}^{N} \Delta I_i = \sum_{i=2}^{N-1} \sigma t V_i + \sigma t \frac{(V_1 + V_N)}{2}$$

$$= \sigma t \left(\sum_{i=2}^{N-1} V_i + \frac{(V_1 + V_N)}{2} \right) \quad (1.168)$$

where N is the number of cells which are adjacent to the zero-volt electrode. With these tools, we are prepared to calculate resistance by the linear equation method. If there are different materials adjacent to the electrode, then $\sigma_{i\text{AVE}}$ must be used in the computation for the ith node.

Example 1.33-1.   Calculate the resistance of the resistor of Example 1.27-1 shown in Fig. 1.24 by using the linear equation method. An equispaced grid is superimposed upon the resistor with nodal voltages on the edges of the resistor as shown in Fig. 1.40.

Equation (1.164) is imposed at all edge nodes, nodes 1, 4, 5, 8, 9, 13, 14, 20, 21, 22, and 23. Equation (1.165) is imposed at the single exterior corner node, node 19. Equation (1.166) is imposed at the single interior corner node, node 12. Equation (1.142) is imposed at all other (interior) nodes. From these conditions, the configuration matrix $[C]$ and the boundary voltage matrix $[BV]$ are formed and the solution for $[NV]$ is obtained from Eq. (1.151).

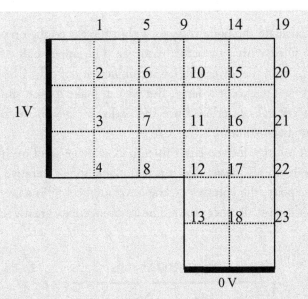

FIGURE 1.40: Grid for linear equation calculation of the resistance.

The resistor current is calculated with $V_{13} = 0.2180$, $V_{18} = 0.1980$, and $V_{23} = 0.1890$ via Eq. (1.161) as $I_R = 0.4014$. Thus, the resistance is given by $R = 2.49/\sigma t\,\Omega$ which compares well with $R = 2.65/\sigma t\,\Omega$ obtained by curvilinear squares in Example 1.27-1.

Example 1.33-2.   Calculate the resistance of the resistor shown in Example 1.33-1 by the iteration method. The equations applied at each node are the same as in Example 1.33-1. The node voltages from an Excel worksheet are shown in Fig. 1.41. The current calculation via Eq. (1.168) gives $I_R = (0.198 + (0.218 + 0.189)/2)\sigma t = 0.4014\sigma t$. Therefore, $R = V_R/I_R = 1/0.4014\sigma t = 2.49/\sigma t\,\Omega$, exactly the result we obtained in Example 1.33-1.

	A	B	C	D	E	F	G
2		1	0.875	0.758	0.659	0.594	0.572
3		1	0.871	0.748	0.642	0.573	0.550
4		1	0.862	0.723	0.587	0.506	0.480
5		1	0.855	0.694	0.476	0.385	0.360
6					0.218	0.198	0.189
7					0	0	0

FIGURE 1.41: Resistor node voltages by the iteration method.

Example 1.33-3. Calculate the node voltages within the closed region of Fig. 1.42(a) that is inhomogeneously filled with conductive material. The upper half of the region has a conductivity of $\sigma = 2$ S/m; the lower half has a conductivity of $\sigma = 1$ S/m. With only nine nodes, the problem is not particularly large. But we can make it even simpler by noting the symmetry about the vertical centerline since the node voltages in the right-hand column equal those in the left-hand column.

We don't need to solve for the latter three node voltages and we have only six nodes for which we must solve. This means that it is reasonable to use iterative hand calculations. Due to the inhomogeneity, the voltages on the horizontal centerline are calculated via Eq. (1.164); Eq. (1.142) is used at all other nodes. The inhomogeneity greatly reduces the accuracy

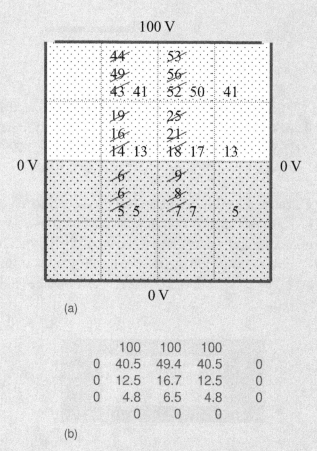

(a)

(b)

FIGURE 1.42: Inhomogeneous material calculations: (a) hand calculations and (b) spreadsheet calculations.

of our previous procedure for estimati ng the initial node voltages. However, since the calculations can begin with any value, we will be close enough if we make our initial estimates as if the media is homogeneous. The initial estimates are the top value at each node; successive iterations are shown in Fig. 1.42(a) with the final values as the last value at each node. The results of spreadsheet iteration are shown in Fig. 1.42(b). The two results show remarkable agreement.

Example 1.33-4. The cylindrical PEC electrodes of a coaxial resistor are located at radii 1 cm and 5 cm; the resistor is 2 cm long. A conductive material between 1 cm and 2 cm has a $\sigma_{\text{INNER}} = 10^3$ S/m; another conductive material between 2 cm and 5 cm has $\sigma_{\text{OUTER}} = 2 \times 10^3$ S/m. Calculate the resistance of this composite structure. Since the equipotentials for a coaxial structure are cylinders of constant radius, the interface is along an equipotential surface so that the two regions act as resistors connected in series. From Example 1.29-1, we know that the resistance of a coaxial resistor is given by $R = \frac{\ln(b/a)}{2\pi\sigma L}$. The combined series resistance is given by

$$R = R_1 + R_2 = \frac{\ln(0.02/0.01)}{2\pi\,10^3(0.02)} + \frac{\ln(0.05/0.02)}{2\pi(2 \times 10^3)(0.02)}$$

$$= 0.0055 + 0.0036 = 9.1 \text{ m}\Omega.$$

Example 1.33-5. A coaxial resistor with dimensions of the resistor of Example 1.33-4 has one conductive material between $0 \leq \phi \leq \pi$ with $\sigma_1 = 10^3$ S/m and another between $\pi \leq \phi \leq 2\pi$ with $\sigma_2 = 2 \times 10^3$ S/m. Calculate the resistance of this coaxial resistor. Since the current flux density points in the radial direction, the material boundaries lie along flux lines and the resistors are in parallel. Each of the resistors has only half the current of a similar coaxial resistor so the resistance of each is doubled. The results of Example 1.29-1 lead to

$$R = (G_1 + G_2)^{-1} = \left(\frac{2\pi\sigma_1 L}{2\ln(b/a)} + \frac{2\pi\sigma_2 L}{2\ln(b/a)} \right)^{-1}$$

$$= \frac{\ln(0.05/0.01)\left(10^3 + 2 \times 10^3\right)^{-1}}{\pi(0.02)} = 8.5 \text{ m}\Omega.$$

1.34 CIRCUIT ANALOGS

An alternative method of calculation of the voltages in a resistor is through analogous behavior of an electric circuit. Consider the four-resistor network of Fig. 1.43. Application of KCL at node O leads to

$$\frac{V_O - V_L}{R} + \frac{V_O - V_R}{R} + \frac{V_O - V_U}{R} + \frac{V_O - V_D}{R} = 0 \qquad (1.169)$$

which, as long as $R \neq 0$, can be rewritten as

$$V_O = \frac{V_L + V_R + V_U + V_D}{4}. \qquad (1.170)$$

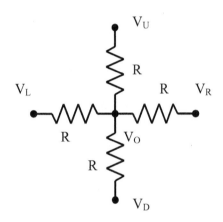

FIGURE 1.43: Analog computation cell.

The relationship between nodes of the electromagnetic problem is the same as that between the nodes of a resistive circuit problem. The current which flows from one node to another is given by Eq. (1.167) which specifies that the analog resistors should have a value of $R = 1/\sigma t$. If only the voltages must be calculated, any convenient value of resistance is acceptable. However, if the actual currents are needed, $R = 1/\sigma t$ must be used. This means that we can simulate the electromagnetic problem by constructing a network of equal-valued resistors. A much quicker method is to use a circuit simulator such as SPICE. The fixed electrode voltages are simulated by voltage sources. Current flow from each electrode is represented by the current from the voltage sources. A brief consideration of the cells of Fig. 1.39 reveals that the resistors between the nodes on the edge of the conducting material only carry half the current of resistors within the material. This is easily achieved by doubling the size of the edge resistors. This solution technique can easily be adapted to three-dimensional structures by substituting a "6" for the "4" in the denominator of Eq. (1.170).

Example 1.34-1. Calculate the node voltages of the closed region of Fig. 1.27 via circuit simulation. For convenience, let's pick $R = 1$ kΩ which is connected between each of the internal nodes. Two sources, 100 and 50 V, represent the electrode voltages. The circuit diagram for this configuration is shown in Fig. 1.44. The PSPICE results are shown in Fig. 1.45. Note that these results compare closely with the exact results of Eq. (1.152).

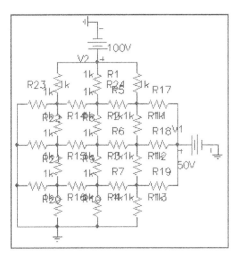

FIGURE 1.44: Circuit diagram for simulation of node voltages.

$$[NV]=\begin{bmatrix} 46.4 \\ 62.1 \\ 64.3 \\ 23.7 \\ 37.5 \\ 45.1 \\ 10.7 \\ 19.2 \\ 28.6 \end{bmatrix} \quad [NV]_{EXACT}=\begin{bmatrix} 46.6 \\ 63.2 \\ 64.8 \\ 23.0 \\ 37.5 \\ 45.2 \\ 10.2 \\ 18.6 \\ 28.4 \end{bmatrix}$$

FIGURE 1.45: Circuit simulation results.

Example 1.34-2. Determine the resistance of the resistor shown in Example 1.33-1 using circuit simulation. Four equal resistors are connected to each of the interior nodes. As with earlier methods, we must impose the condition that no current leaves the edges of the conductive material. This is easily accomplished by the absence of a resistor to carry current away from the edge. In addition, the tangential current in the edge cells is only half of the current of interior cells as expressed by Eq. (1.164), see Fig. 1.39. The current is halved by making the resistors along the edges twice the value of all other resistors, i.e., $R_E = 2R$. With a 1 V source, the total resistance is the reciprocal of the current flowing from the source scaled by the factor $1/\sigma t$. The PSPICE circuit diagram is shown in Fig. 1.46. The current in the source is $I_{Source} = 0.4014$ mA giving a value of $R = 2.49/\sigma t$ which compares very well with our earlier calculations.

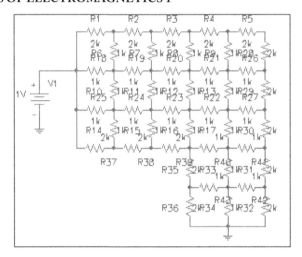

FIGURE 1.46: SPICE circuit diagram for resistance calculation.

1.35   POWER DISSIPATION

The application of a voltage drop across a resistor causes an average power dissipation expressed as

$$P_{\text{DISS}} = \frac{V_R^2}{R} = I_R^2 R \qquad (1.171)$$

where V_R and I_R are the RMS values of the voltage and current, respectively. But, how is this expressed in terms of the electromagnetic fields within the resistor? Consider the incremental resistor shown in Fig. 1.46. The incremental voltage drop is expressed in terms of the electric field as

$$\Delta V = |\mathbf{E}|\Delta l. \qquad (1.172)$$

The incremental current flow can be expressed in terms of the electric field, also, as

$$\Delta I = |\mathbf{J}|\Delta a = \sigma |\mathbf{E}|\Delta a. \qquad (1.173)$$

Assuming that the voltage and current are in phase (as they are for DC voltage and current), we obtain the average incremental power dissipated as

$$\Delta P_{\text{DISS}} = \Delta V \Delta I = \sigma |\mathbf{E}|^2 \Delta a \Delta l = \sigma |\mathbf{E}|^2 \Delta \forall \qquad (1.174)$$

where $\Delta \forall$ is the volume of the incremental resistor. The power dissipated is related to the fields as we suspected. Division of the equation by the volume leads to

$$p_{\text{DISS}} = \frac{\Delta P_{\text{DISS}}}{\Delta \forall} = \sigma |\mathbf{E}|^2, \qquad (1.175)$$

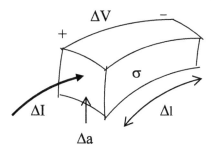

FIGURE 1.47: Increamental resistor.

the power dissipated per unit volume also known as the *power density* with units of w/m³. The power dissipated at each point throughout the resistor is proportional to $|\mathbf{E}|^2$. The locations where $|\mathbf{E}|$ is the greatest dissipate the most power per unit volume. The total power is expressed as the sum of all the incremental powers dissipated throughout the resistor, or

$$P = \sum_{\text{RESISTOR}} \Delta P_{\text{DISS}} = \sum_{i=1}^{N} p_{\text{DISS}_i} \Delta v_i$$

$$= \sum_{i=1}^{N} \sigma_i |\mathbf{E}_i|^2 \Delta v_i \tag{1.176}$$

where the volume of the resistor has been divided into N subvolumes. In the limit, as the subvolumes all shrink to zero, the sum is replaced by an integral expression as

$$P = \iiint_\forall \sigma |\mathbf{E}|^2 dv. \tag{1.177}$$

Note that both σ and $\mathbf{E}$ can vary from point to point; power density—the localized power dissipation—varies as well. Once $\mathbf{E}$ has been determined, the total power dissipated throughout a region can be calculated by summing (integrating) the incremental (differential) power dissipated throughout the resistor.

Example 1.35-1. Calculate the total power dissipated in the coaxial resistor of Example 1.29-1. $\mathbf{E}$, given by Eq. (1.177), becomes

$$\mathbf{E} = -\nabla V = -\frac{\partial V}{\partial \rho}\mathbf{a}_\rho = -\frac{V_O}{\ln(a/b)\rho}\mathbf{a}_\rho = \frac{V_O}{\ln(b/a)\rho}\mathbf{a}_\rho,$$

so that

$$P = \iiint_\forall \sigma |\mathbf{E}|^2 dv = \int_{z=0}^{L} \int_{\phi=0}^{2\pi} \int_{\rho=a}^{b} \frac{\sigma V_O^2}{(\ln (b/a))^2 \rho^2} \rho d\rho d\phi dz$$

$$= \frac{2\pi\sigma L V_O^2}{\ln(b/a)}.$$

From circuit theory, the power is calculated as $P = \frac{V_R^2}{R} = \frac{V_O^2}{\left(\frac{\ln(b/a)}{2\pi\sigma L}\right)} = \frac{2\pi\sigma L V_O^2}{\ln(b/a)}$ in agreement with the field-based calculation.

The common circuit definition of resistance, $R = V_R/I_R$, can be expressed in terms of $\mathbf{E}$ as

$$R = \frac{V_R}{I_R} = \frac{\int_L \mathbf{E} \cdot \mathbf{dl}}{\iint_S \mathbf{J} \cdot \mathbf{ds}} = \frac{\int_L \mathbf{E} \cdot \mathbf{dl}}{\sigma \iint_S \mathbf{E} \cdot \mathbf{ds}}. \tag{1.178}$$

It has been suggested that power dissipation is a more fundamental concept upon which to base the definition of resistance. This leads to an expression for resistance of

$$R = \frac{P}{I_R^2} = \frac{\iiint_\forall \sigma |\mathbf{E}|^2 dv}{\left(\iint_S \mathbf{J} \cdot \mathbf{ds}\right)^2} = \frac{\iiint_\forall |\mathbf{E}|^2 dv}{\sigma \left(\iint_S \mathbf{E} \cdot \mathbf{ds}\right)^2}. \tag{1.179}$$

Alternatively, the power dissipation definition can be expressed as

$$R = \frac{V_R^2}{P} = \frac{\left(\int_L \mathbf{E} \cdot \mathbf{dl}\right)^2}{\sigma \iiint_\forall |\mathbf{E}|^2 dv}. \tag{1.180}$$

As seen in all three of these forms, once $\mathbf{E}$ is known the resistance can be calculated. The latter two forms are particularly useful for optimization techniques. But these methods are beyond the scope of this course. Suffice it to say that these optimization procedures are based upon the fact that the fields and current within a resistor naturally distribute themselves so as to minimize the power dissipated. Any other distribution causes greater power dissipation.

At this point, we have covered the basic electromagnetics which govern the behavior of resistors. More advanced topics have been deferred to more advanced courses. We will now move on to the behavior of capacitors.

CHAPTER 2

Capacitors

2.1   CAPACITORS: A FIRST GLANCE

The basic function of a capacitor is as a storage element for electric energy. Its configuration is designed to enhance this function. Properly selected materials minimize its power dissipation as well. As you recall from circuits, the terminal behavior of an ideal capacitor is given by

$$I_C = C \frac{dV_C}{dt} \qquad (2.1)$$

where I_C is the current flowing into the capacitor in the direction of the voltage drop across the capacitor, V_C, and C is the capacitance value of the capacitor expressed in Farads and abbreviated as F. Its terminal behavior is more complicated than that of a resistor due to the presence of the derivative of voltage. As with resistors, we will investigate the internal, electromagnetic behavior of capacitors to better understand them.

The simplest capacitor configuration has many similarities with that of resistors, see Fig. 2.1. Two metallic wire leads provide the connection between a capacitor and the external circuit. Current enters the element at one end through the wire lead and flows directly onto a metallic electrode. An equal current flows from the other electrode out of the capacitor via the other wire lead. The flux guiding material is located between the two metal electrodes. An insulator serves as the electric flux guide in a capacitor similar to the way conductive material in a resistor guides the current flux. The charges that enter the capacitor do not flow to the opposite electrode and leave via the other lead because the conductivity of the insulator is zero. Instead, they accumulate on the electrode while an equal charge flows from the other electrode out of the other wire lead. This leaves equal but opposite charges on the two electrodes. The use of an insulating flux guide instead of one that is conductive is the chief reason for the marked difference in behavior between capacitors and resistors.

As with resistors, the electrode-material configuration of a capacitor greatly affects its element value. The flux guiding material is often chosen to minimize losses and to enhance the capacitor's energy storage capability as well as its tendency to resist arcing. Usually the entire assembly is hermetically encapsulated to minimize environmental effects on the capacitor's behavior.

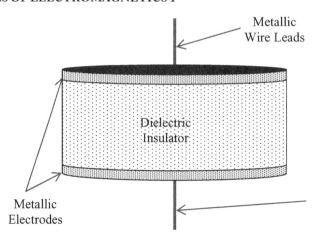

FIGURE 2.1: Physical configuration of a capacitor.

The series resistance offered by the metallic wire leads and electrodes is extremely small and produces a negligible voltage drop in most capacitors. This behavior is most simply modeled by approximating the capacitor leads and electrodes as PECs with no resistance and no voltage drop. The extremely small conductivity of most insulators (see Appendix D) is modeled by approximating the conductivity of dielectric insulators as zero. A further simplification is achieved if the flux guide is assumed to be ideal, i.e., no flux leaks out of the dielectric insulator. This closely models the behavior of many capacitors, especially those for which the spacing between electrodes is very small compared to their planar dimensions. Furthermore, this approximation allows us to employ many of the techniques that we used in the analysis of resistors in Chapter 1. Finally, we will use the same simple configuration for capacitors that we used for resistors. These approximations are realized in the electrical and configuration models shown in Fig. 2.2. Two identical, planar PECs of surface area A form the electrodes of the capacitor model. The electrodes are parallel to each other and separated by a height h which is much less than either of the planar dimensions of the electrodes. An ideal dielectric material occupies the region between the two electrodes. This model is known as a parallel electrode capacitor.

2.2   CHARGES ON ELECTRODES

Conservation of charge requires that the net charge within a volume changes only when current flows into or out of the volume. This is expressed by Eq. (1.47) as

$$I_{\text{OUT}} = \oiint_S \mathbf{J} \cdot \mathbf{ds} = -\frac{dQ}{dt}. \tag{2.2}$$

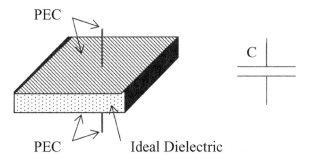

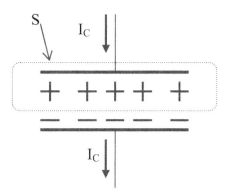

FIGURE 2.2: Models for capacitors: (a) physical model and (b) electrical model.

Its application to a capacitor is illustrated in Fig. 2.3. The capacitor wire leads carry electric charges, the only currents that enter or leave the capacitor, through the surface S which encloses the entire capacitor. Our experience with circuits shows that the current entering the one lead of a capacitor is equal to the current leaving the other; this means that $I_{OUT} = 0$. If current flows into one lead and out of the other lead of a capacitor, then there is no change in the charge within the capacitor, i.e., $dQ_C/dt = 0$. Though there is no net accumulation of charge within the capacitor, charge cannot flow from one electrode to another due to the presence of an insulator between the electrodes. Consequently, all of the charge that

FIGURE 2.3: Charge conservation in a capacitor.

flows into one lead of the capacitor accumulates on that electrode. Simultaneously, an equal current flows out of the other electrode depleting the charge on it by the same amount as the accumulation on the other electrode. The charges on the two electrodes are equal, but opposite in sign so that their sum is zero and there is no net charge on the capacitor.

The charge on an electrode of the capacitor is obtained by integrating Eq. (2.2). For the current entering the positive lead of a capacitor, I_C, this gives

$$Q(t) - Q(t_o) = \int_{t'=t_o}^{t} I_C(t')dt' \ [\text{C}]. \qquad (2.3)$$

From the charge conservation principle, the positive sign for the RHS occurs because the current entering the capacitor is negative outward current; this negative sign cancels the negative sign of Eq. (2.2). From circuits we know that a positive current into the positive node results in positive charge accumulating on this electrode; Eq. (2.3) expresses this. But, how is this charge distributed on an electrode?

Experimental evidence has shown that excess mobile charges are not contained within the interior of a metal object but they reside on its surface. This is mainly due to the fact that like charges tend to repel each other; a surface distribution allows them to be as far apart as possible and to minimize their collective energy. A detailed description of this distribution depends upon the geometry of the metal object. The mathematical challenge of calculating a surface charge distribution will be delayed for a while. The presence of a second electrode affects the charge distribution, also. The mutual attraction of the equal and opposite charges on the two electrodes tends to concentrate them on the inner electrode surfaces where they are nearest to each other with little or no charge on the outside electrode surfaces.

Let's apply these fundamental ideas regarding charge distribution to the parallel plate capacitor in Fig. 2.2. As a first approximation, assume that the charge is distributed uniformly over the inner surfaces of the electrodes. Any charges that tended to reside on the outer surfaces have been "pulled" to the inner surfaces by the attractive forces of the opposite charges on the other electrode. Experimental evidence shows that this approximation is reasonably accurate (except near the electrode edges) for capacitors when the height h is relatively small compared to the planar dimensions of the electrodes.

With this charge distribution model, we can picture the charge entering the capacitor through the PEC wire lead and nearly instantly* spreading into a uniform distribution over the inner surfaces of the PEC electrodes. The uniform surface charge density is calculated as

$$\rho_S(t) = \frac{Q(t)}{A} = \frac{\int_{t'=t_o}^{t'=t} I_C(t')dt'}{A} \ [\text{C/m}^2] \tag{2.4}$$

where $Q(t_o) = 0$ is assumed for simplicity. The surface charge density is uniform and time varying.

Example 2.2-1. The current flowing into a capacitor is given by $I_C(t) = 10 \cos 377t$ A. The electrodes both have dimensions of 1 cm $\times$ 1 cm. Express the charge on the electrodes as a function of time. Assuming uniform charge distribution, express the surface charge density on the positive electrode as a function of time. From Eq. (2.3), the charge on the positive electrode is expressed as $Q(t) = Q(t_o) + \int_{t'=t_o}^t I_C(t')dt' = \int_{t'=0}^t I_C(t')dt'$ where the initial charge on the capacitor at $t = t_o = 0$ is assumed to be zero. We find $Q(t) = (10/377)\sin 377t$ C on the

---

* No, the charges do not move faster than the speed of light. But for capacitors used in circuits, the time required for the charges to become uniformly distributed is orders of magnitude smaller than circuit time constants. So we treat it as instantaneous.

positive electrode and the opposite on the negative electrode. The surface charge density on the positive electrode is calculated from Eq. (2.4) as $\rho_S(t) = Q(t)/A = (10^5/377)\sin 377t$ C/m^2.

2.3   GAUSS' LAW

The development of the theory of electric charges and the fields surrounding them involved many physicists and mathematicians. Most notable was an intricate series of experiments conducted by Klaus Friedrich Gauss that showed that the behavior of charges can be described in terms of an *electric flux density* which emanates from all charges. Gauss found that the total electric flux emanating from a charge is equal to the magnitude of the charge

$$\Psi_e = Q \ [\text{C}].\tag{2.5}$$

This electric flux is not representative of any actual motion. Instead, it can be thought of as fixed lines of flux projecting from the charge and occupying the space around it. The fixed gravitational flux that emanates from objects with mass is analogous. Gauss introduced the idea of an electric flux density, **D**, with units of C/m^2, emanates from each element of charge. The total flux emanating from a charge can be expressed as

$$\Psi_e = \oiint_S \mathbf{D} \cdot \mathbf{ds} = Q.\tag{2.6}$$

The electric flux passing through any closed surface is equal to the charge enclosed. This powerful relationship, known as Gauss' law, provides the connection between charge and total electric flux. The closed surface is often called a *Gaussian surface* to emphasize this important principle. We can interpret the meaning of Gauss' law and apply the evaluation techniques to it in the same manner as we did with flux integrals of current density.

Application of Gauss' law can be seen with the various charges and surfaces within Fig. 2.4. Point charges Q_A, Q_B, and Q_C are the only charges contained within surface S_1. Therefore, the total electric flux through the closed surface S_1 is given by

$$\Psi_1 = Q_A + Q_B + Q_C.\tag{2.7}$$

The closed surface S_2 contains a volume charge density and point charge Q_B, so the total flux through surface S_2 is given by

$$\Psi_2 = Q_B + \iiint_\forall dv.\tag{2.8}$$

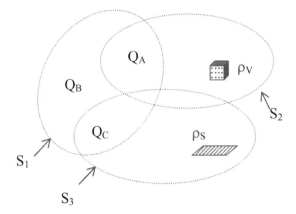

FIGURE 2.4: Application of Gauss' law.

Surface S_3 contains a surface charge density and point charge Q_C, so the total flux through surface S_3 is given by

$$\Psi_3 = Q_C + \iint_A \rho_s da. \tag{2.9}$$

But, keep in mind that when the total flux out of a surface is zero, it does not mean that there is no charge contained within the surface. Rather, it means that the sum of the charges is zero and that at some points on the surface electric flux is outward directed and at other points it is inward directed. The algebraic sum of all the fluxes is zero. Since we do not know the electric flux density, we cannot evaluate the surface integral for these charge distributions. But, we are able to use the enclosed charge as a means of calculating the total flux emanating from a particular surface.

Example 2.3-1. A spherical volume centered on the origin and with a radius $r = 10\,\text{m}$ contains a volume charge density of $\rho_V = (0.1/r)\ \text{C/m}^3$ and a point charge of $-10\,\text{C}$. Calculate the total electric flux emanating from the sphere. Gauss' law states that the total electric flux emanating from a closed surface is the sum of the charges enclosed by the surface, $\Psi_e = Q$. Consequently,

$$\Psi_e = Q = -10 + \iiint_{r=10} \frac{0.1}{r} dv = -10 + 0.1 \int\limits_{r=0}^{10} \int\limits_{\theta=0}^{\pi} \int\limits_{\phi=0}^{2\pi} \frac{r^2 \sin\theta}{r} d\phi d\theta dr$$

$$= -10 + 0.1(4\pi)(50) = 20\pi - 10\,\text{C}.$$

Note that even though the volume charge density becomes unbounded at $r = 0$ due to $1/r$, the differential charge remains finite since $dQ = r \sin\theta \, dr \, d\theta \, d\phi$.

Example 2.3-2. Calculate the electric flux that emanates from a closed surface that surrounds the less positive electrode of the capacitor of Example 2.2-1. The electric flux through any closed surface surrounding the negative electrode is equal to the charge on it, $\Psi_e = Q(t) = -(10/377) \sin 377t$ C. Note that Ψ_e varies with time.

2.4 DIVERGENCE OF D

The pointwise behavior of **D** is expressed through the divergence just as it was for **J**. Since charge never accumulates or depletes anywhere within a resistor, the divergence of **J** vanishes everywhere, i.e., $\nabla \cdot \mathbf{J} = 0$. However, since the net electric flux out of a volume that contains charge is not zero, we should expect that $\nabla \cdot \mathbf{D}$ may behave differently therein.

Beginning with Gauss' law,

$$\Psi_e = \oiint_S \mathbf{D} \cdot \mathbf{ds} = Q, \tag{2.10}$$

we evaluate the LHS for an incrementally small volume, following the procedure used for **J** in Section 1.19, to obtain

$$\Delta\Psi_e \approx (\nabla \cdot \mathbf{D})\Delta\mathsf{V}. \tag{2.11}$$

If we consider all the charge within surface S to be volume-distributed charge, ρ_V, then as the volume Δv contained within S becomes incrementally small the RHS of Eq. (2.6) becomes

$$Q = \iiint_\mathsf{V} \rho_V dv \approx \rho_V \Delta\mathsf{V}. \tag{2.12}$$

Since Eqs. (2.11) and (2.12) are equal, the divergence of the electric flux density is given by

$$\nabla \cdot \mathbf{D} = \rho_V. \tag{2.13}$$

This is the point form of Gauss' law. $\nabla \cdot \mathbf{D}$ is not identically zero since electric flux lines begin and end upon charge. The existence of ρ_V at a point necessitates that **D** diverge there and $\nabla \cdot \mathbf{D} \neq 0$.

This form is not valid for surface or line charge densities, i.e., ρ_S and ρ_L, respectively, since they do not vary continuously throughout space, as does ρ_V. The complication for these distributions occurs in the evaluation of the RHS of Eq. (2.5), which does not contain a Δv term. However, the integral form of Gauss' law is always valid for any charge distribution.

Of course, the divergence theorem of calculus,

$$\Psi_e = \oiint_S \mathbf{D} \cdot \mathbf{ds} = \iiint_\forall \nabla \cdot \mathbf{D} dv = Q, \qquad (2.14)$$

applies in all regions where the derivatives of $\nabla \cdot \mathbf{D}$ are finite. This provides an alternative integral for evaluating the charge contained within a surface S.

Due to the tight atom–electron bonds, there is rarely any free charge within dielectrics and so $\nabla \cdot \mathbf{D} = 0$ therein. However, free charge on the surface is much more likely. Once charge exists on a dielectric surface, the near-zero value of conductivity prevents it from moving any significant distance. Dealing with this phenomenon is discussed later with boundary conditions.

Example 2.4-1.   Within a region of space, the electric flux density is given as $\mathbf{D} = r\mathbf{a}_r = x\mathbf{a}_X + y\mathbf{a}_Y + z\mathbf{a}_Z$ C/m^2. Calculate the electric flux emanating from a unit cube centered on the coordinate origin. Calculation of electric flux is defined as the surface integral of flux density over the surface. For this problem, there are six surfaces located at $x = \pm 0.5$, $y = \pm 0.5$, and $z = \pm 0.5$. The integrals are relatively easy and all evaluate as

$$\Psi_{\text{SIDE}} = \iint_A D_{\text{SIDE}} da = \int_{z=-0.5}^{0.5} \int_{y=-0.5}^{0.5} 0.5 \, dy dz = 0.5 \, \text{C}$$

so that the total flux is $\Psi_e = 6\Psi_{\text{SIDE}} = 3$ C. Alternatively, the flux can be calculated by the divergence theorem. First calculate $\nabla \cdot \mathbf{D} = 3$. Then integrate throughout the volume of the unit cube to obtain $\Psi_e = 3$ C as before. Frequently, though not always, the volume integral is easier than the surface integral.

2.5   DIELECTRIC PERMITTIVITY

The parameter of an insulator that characterizes its dielectric properties is known as *permittivity*. It is denoted by the symbol ε and has units of Farads/meter (F/m) in the SI system. Permittivity in dielectric material serves a role similar to the conductivity in conductive material; it relates the field intensity to the flux density.

Earlier in this chapter we found that $\mathbf{D}$, a flux density, is present between the two electrodes of a capacitor when they have equal and opposite charges on them. But the presence of these charges also creates $\mathbf{E}$, a field intensity. In a resistor, we found that the current flux density and the electric field intensity were related by the conductivity of the material. In a similar fashion, experiments have shown that the electric flux density and the electric field intensity are related by the permittivity property of the material as

$$\mathbf{D} = \varepsilon\mathbf{E} \qquad (2.15)$$

throughout the dielectric material. In both cases, the material *constitutive* property relates the flux density to the field intensity even though quite different phenomena are involved. We shall later see that a similar constitutive relationship holds for magnetic fields as well.

To describe the cause and nature of permittivity, we will model the dielectric material as a lattice of neutral atoms. A positive nucleus at the center of the atom is encircled by a "cloud" of negatively charged electrons; the sum of the charges is zero. The atoms have a very "tight" grip on their electrons. In fact, in an ideal dielectric, electrons cannot break away from the central atom. Note how this contrasts with weak bond between atoms and electrons in conductors. In PEC material, the atoms' grip on electrons approaches zero so that current flow occurs without a voltage drop. In the absence of an external electric field, the center of positive charge of a dielectric atom is aligned with the center of negative charge, see Fig. 2.5(a). But, in the presence of an applied electric field, Coulomb forces displace the positive nucleus ever so slightly in the direction of the electric field; the center of the negative "cloud" is displaced slightly in the other direction, see Fig. 2.5(b). This displacement polarizes the atom. Though the displacements are extremely small, on the order of 10^{-17} m for typical dielectrics, the total charge shifted is significant due to the large number of atoms, on the order of 10^{23} atoms per cubic meter. The displacement converts electrical energy of the electric field into stored mechanical energy within each atom, much like a spring. With the removal of the electric field, the charge centers are aligned again and the potential energy is returned to the electric field without any energy loss. The displacement of the charge centers creates an internal electric field directed from the positive charge center to the negative and in opposition to the applied external field. The strength of the opposing field is related to the dielectric properties. The total field between the

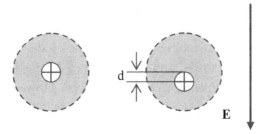

FIGURE 2.5: Effects of external electric field on adielectric atom: (a) without field and (b) with field.

electrodes, i.e., within the dielectric, is reduced by this opposing field. This opposing field is called the *polarization field*.

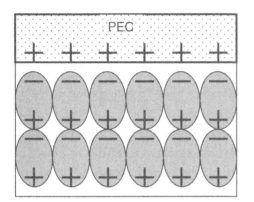

FIGURE 2.6: Polarization charges at the dielectric—electrode interface.

When there is a vacuum between the electrodes, i.e., no dielectric, there is no polarization field and the field between the electrodes is equal to the external field. When a dielectric is present, the field between the electrodes is reduced by the presence of the dielectric. This may be explained by looking at the dielectric material where it contacts the PEC electrodes, see Fig. 2.6. Though the charge displacement produces no excess charge within the material, at the boundary with the positive electrode there is a thin layer of negative charge and at the negative electrode, a positive layer of charge. These charge layers on the surface of the dielectric attract an equal charge of the electrode to their vicinity. The dielectric polarization charges cannot combine with the free surface charges of the electrode since the dielectric charges are bound to their central atoms. However, since they are in such close proximity, their electric fields cancel, thus reducing the internal electric field within the dielectric material.

The permittivity of free space, e.g., a vacuum, is noted as ε_o and has a value of $\varepsilon_o = 8.854 \times 10^{-12}$ F/m in the SI system. The permittivity of all, but a very few, dielectrics is greater than that of free space. The common notation to describe the permittivity of dielectric materials is *relative permittivity*,

$$\varepsilon_R = \frac{\varepsilon}{\varepsilon_o} \qquad (2.16)$$

where ε_R is the ratio of the dielectric permittivity to free space permittivity and is usually greater than 1. Frequently, ε_R is called *dielectric constant*. Most insulators have dielectric constants in the range $1 < \varepsilon_R < 10$. The dielectric constants for common dielectric materials are given in Appendix D.

There are two distinct shortcomings to this simple model for dielectrics. Firstly, the previous discussion was confined to DC and low-frequency AC fields. As the signal frequency is increased, the permittivity shows resonance and loss effects. However, these shortcomings do not invalidate the primary dielectric property of energy storage. We will leave these more subtle topics for an advanced course. Secondly, we have ignored effects which "strip away" electrons from their central atom. But, in fact, every material has a critical value of field strength at which electrons can be stripped away from the parent atom. Usually, this is accompanied by an arc and with sudden, catastrophic current flow. This process is known as *dielectric breakdown*. The field strength at which this phenomenon occurs is known by several names, *critical field strength*, *dielectric breakdown strength*, and *dielectric strength*, and is denoted by E_C. For air at sea level, E_C is on the order of 3×10^6 V/m.

2.6 DIELECTRIC BOUNDARY CONDITIONS

As we found with conductors, the fields on opposite sides of material boundary are related to each other through the properties of the two materials. The development of this relationship proceeds much as with conductive materials; see Section 1.25.

The conservative nature of the electric field is used again to obtain the tangential boundary condition as obtained earlier as

$$E_{T1} = E_{T2} \tag{2.17}$$

and in vector form

$$\mathbf{a_N} \times (\mathbf{E}_1 - \mathbf{E}_2) = 0. \tag{2.18}$$

The tangential components of the electric field on both sides of a boundary are the same.

The electric flux density **D** in dielectrics behaves much as **J** in resistors where free charges are not present. Free charge rarely exists within dielectric material, but is much more common as surface charge density at a material interface. Such a situation is shown in Fig. 2.7. As we did in establishing the boundary conditions on **J**, we center a short box centered on the boundary so that the surface charge density remains within the box as its height approaches zero.

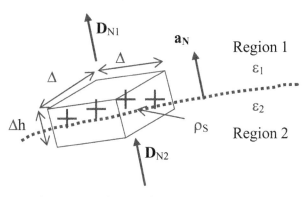

FIGURE 2.7: Electric flux density boundary conditions.

Application of Gauss' law and the methods of Section 1.25 indicate that flux only emanates from the top and bottom of the box. Analogous to Eq. (1.97), we obtain the electric flux out of the box as

$$\oiint_{\text{BOX}} \mathbf{D} \cdot \mathbf{ds} \approx (\mathbf{D}_{\text{TOP}} - \mathbf{D}_{\text{BOTTOM}}) \cdot \mathbf{a}_N \Delta w \Delta w$$

$$= (D_{N2} - D_{N1}) \Delta w \Delta w, \tag{2.19}$$

which by Gauss' law must equal the charge contained within the box. As the height of the box shrinks to zero, the only charge remaining within the box is the surface charge on the boundary given as

$$Q = \int\int_{\Delta w \, \Delta w} \rho_s da \approx \rho_s \Delta w \Delta w. \tag{2.20}$$

Setting Eqs. (2.19) and (2.20) equal, we obtain the boundary conditions on $\mathbf{D}$ as

$$D_{N2} - D_{N1} = \rho_S. \tag{2.21}$$

You may see this expressed in a vector format as

$$\mathbf{a_N} \cdot (\mathbf{D}_2 - \mathbf{D}_1) = \rho_S \tag{2.22}$$

where $\mathbf{a_N}$ is directed from region 2 to region 1. The normal components of the electric flux density, $\mathbf{D}$, at a boundary are discontinuous by the surface charge density, ρ_S, on the surface.

The occurrence of free surface charge on dielectrics is rare; for most dielectric–dielectric boundaries $\rho_S = 0$. When $\rho_S = 0$, we can combine Eqs. (2.17), (2.21), and $\mathbf{D} = \varepsilon \mathbf{E}$ to express both boundary conditions for dielectric–dielectric boundaries either in terms of $\mathbf{E}$ or in terms of $\mathbf{D}$ as

$$E_{T1} = E_{T2} \quad \text{and} \quad \varepsilon_1 E_{N1} = \varepsilon_2 E_{N2}$$
$$\frac{D_{T1}}{\varepsilon_1} = \frac{D_{T2}}{\varepsilon_2} \quad \text{and} \quad D_{N1} = D_{N2}. \tag{2.23}$$

We have already seen that a surface charge density resides on the facing electrode surfaces within a capacitor. When region 2 of Fig. 2.5 is a PEC electrode, $\mathbf{E}_2 = 0$ within a PEC so that $\mathbf{D}_2 = \varepsilon_2 \mathbf{E}_2 = 0$ and $D_{N2} = 0$. Consequently, the boundary conditions for a dielectric–PEC interface become

$$D_{T1} = 0 \quad \text{and} \quad D_{N1} = \rho_S \qquad (2.24)$$

where ρ_S is the surface charge density on the electrode. Through the boundary conditions, we have now established the relationship between the charge density on the electrodes and the electric flux density at the dielectric–electrode boundary.

Since $\mathbf{E}_2 = 0$ within the PEC electrode, $E_{T2} = 0$. Boundary conditions require that $E_{T1} = E_{T2} = 0$ so that $D_{T1} = \varepsilon_1 E_{T1} = 0$. This means that in the dielectric material the $\mathbf{E}_1 = \mathbf{a}_N E_N$ and $\mathbf{D}_1 = \mathbf{a}_N \varepsilon_1 E_N$, i.e., the electric field and flux density are perpendicular to the PEC electrode as they are in resistors. Moreover, note that they are directed in the direction of $\mathbf{a}_N$ away from the positive surface charge.

As you recall from Section 1.25, the amount of current flux which "leaked" out of the conductive material was proportional to the ratio of the conductivities of air and of the resistor. The smaller this ratio, the more effectively the resistor functions as an ideal flux guide. The same principle applies to dielectric material. The phenomenon of flux leaking from the dielectric of a capacitor into air is known as *fringing*. The smaller the fringing field, the more accurate are the assumptions that all the flux is contained within the dielectric. From Eq. (2.21), we see that $D_{T2} = (\varepsilon_2/\varepsilon_1)D_{T1}$. When the dielectric of the capacitor is taken as region 1 and the surrounding air as region 2, we see that if the ratio $\varepsilon_2/\varepsilon_1 = 1/\varepsilon_R < 0.1$, then the fringing field is so much smaller than the dielectric field that it can be neglected. For a good flux guide, the larger ε_R, the better. Unfortunately, $1 < \varepsilon_R < 10$ for most dielectrics, so this condition is not always valid. Nevertheless, we will assume no fringing fields for most capacitors. The significance of the fringing fields is reduced when the spacing of the plates is relatively small compared to the largest dimension of the electrodes. This condition is often met and strengthens our "no fringing" assumption.

Example 2.6-1.   A dielectric material with relative permittivity of $\varepsilon_R = 2$ occupies the region for $x \leq 0$ and is bounded by air at the $x = 0$ plane. The electric field within the dielectric at the boundary is given by $\mathbf{E} = 5\mathbf{a}_X + 5\mathbf{a}_Y - 5\mathbf{a}_Z$ V/m. There is no surface charge density upon the boundary. Calculate $\mathbf{E}$ and $\mathbf{D}$ in the air adjacent to the boundary. If the normal to the boundary is defined as $\mathbf{a}_N = \mathbf{a}_X$, then the dielectric region represents region 2 in Fig. 2.7 and $E_{N2} = 5$ V/m and $E_{T2} = 5\sqrt{2}$ V/m with a direction of $\mathbf{a}_T = 0.707(\mathbf{a}_Y - \mathbf{a}_Z)$. In addition $D_{N2} = 2\varepsilon_o 5 = 10\varepsilon_o$ and $D_{T2} = 10\sqrt{2}\varepsilon_o$ C/m². From Eq. (1.17), we obtain $E_{T1} = E_{T2} = 5\sqrt{2}$ V/m with $D_{T1} = \varepsilon_o E_{T1} = 5\sqrt{2}\varepsilon_o$ C/m² and $D_{N1} = D_{N2} = 10\varepsilon_o$ C/m² and $E_{N1} = D_{N1}/\varepsilon_o = 10$ V/m. Thus the fields in air are $\mathbf{E}_1 = 10\mathbf{a}_X + 5\sqrt{2}(0.707)$

$(\mathbf{a}_Y - \mathbf{a}_Z) = 10\mathbf{a}_X + 5(\mathbf{a}_Y + \mathbf{a}_Z)$ V/m and $\mathbf{D}_1 = \varepsilon_o \mathbf{E}_1 = 10\varepsilon_o \mathbf{a}_X + 5\sqrt{2}\varepsilon_o \mathbf{a}_T = 10\varepsilon_o \mathbf{a}_X + 5\varepsilon_o(\mathbf{a}_Y - \mathbf{a}_Z)$ C/m^2.

Example 2.6-2.   Consider Example 2.6-1 with the addition of a surface charge density of $\rho_S = -5\varepsilon_o$ C/m^2 on the dielectric–air interface. The tangential components are not affected by surface charge density. The normal components satisfy Eq. (2.21) so that $D_{N1} = D_{N2} + \rho_S = 10\varepsilon_o - 5\varepsilon_o = 5\varepsilon_o$ C/m^2. The resultant fields are $\mathbf{E}_1 = 5(\mathbf{a}_X + \mathbf{a}_Y + \mathbf{a}_Z)$ V/m and $\mathbf{D}_1 = 5\varepsilon_o(\mathbf{a}_X + \mathbf{a}_Y + \mathbf{a}_Z)$ C/m^2. Note that the addition of surface charge density alters only the normal components of the fields.

2.7   FLUX TUBES, EQUIPOTENTIALS, AND CAPACITANCE

Gauss' law predicts that electric flux density, $\mathbf{D}$, emanates perpendicularly from surface charge, ρ_S, on the PEC electrodes of a capacitor into the dielectric material. Since there are no free charges within an ideal dielectric, $\nabla \cdot \mathbf{D} = 0$ and $\mathbf{D}$ lines begin and end only on the electrodes. Moreover, $\mathbf{D}$ lines coincide with $\mathbf{E}$ lines and are always perpendicular to equipotentials. Consequently, electric flux is confined to tubes; flux never leaks from nor enters a flux tube. This model of the fields in capacitors is remarkably similar to resistors. Flux tubes begin and end on the electrodes and extend from one electrode to the other. Equipotentials are perpendicular to the flux tubes, see Fig. 2.8.

Up to this point, we have determined the following features of the fields within a capacitor:

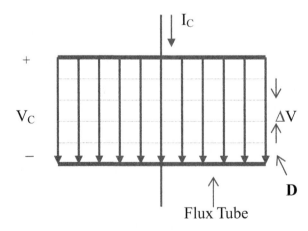

FIGURE 2.8: Flux tubes and equipotentials in a capacitor.

(1) Current flow, I_C, into the PEC lead deposits charge, $+Q$, on one electrode. Equal current flow out of the other PEC lead leaves a $-Q$ on the other electrode. The net charge in the capacitor is zero.

(2) These charges, $\pm Q$, are assumed to be uniformly distributed on the facing surfaces of the electrodes with a surface charge density of $\rho_S = Q/A$.

(3) Electric flux density **D** begins and ends on the electrode surface charge density where there is only a normal component $D_N = \rho_S$.

(4) Flux tubes of **D** extend within the insulating dielectric from one electrode to the other.

(5) **D** is perpendicular to surfaces of constant V.

(6) Electric flux density is related to the electric field intensity by $\mathbf{D} = \varepsilon\mathbf{E}$, where ε is the permittivity of the dielectric material.

(7) The voltage drop between two points is defined as $V_{\text{FINAL}} - V_{\text{INITIAL}} = -\int_L \mathbf{E} \cdot d\boldsymbol{\ell}$.

These concepts are now applied to obtain an expression for capacitance.

To obtain an expression for capacitance from Eq. (2.1),

$$I_C = C\frac{dV_C}{dt} \qquad (2.25)$$

the terminal voltage and current of a capacitor are expressed in terms of the internal fields. In Chapter 1, we found that the equipotentials for parallel electrodes of resistors are parallel to the electrodes. Due to the similar geometry, this is true for parallel electrode capacitors with closely spaced electrodes as well. For a voltage drop of V_c, as shown in Fig. 2.8, the voltage within the capacitor is expressed as

$$V(z) = V_C\frac{z}{h}, \qquad (2.26)$$

with an accompanying electric field of

$$\mathbf{E} = -\frac{V_C}{h}\mathbf{a}_Z. \qquad (2.27)$$

Substituting this into Eq. (2.15), we obtain

$$\mathbf{D} = D_Z\mathbf{a}_Z = \varepsilon E_Z\mathbf{a}_Z = -\frac{\varepsilon V_C}{h}\mathbf{a}_Z. \qquad (2.28)$$

This result leads to an expression for the terminal voltage in terms of the flux density and the physical parameters of the capacitor as

$$V_C = -\frac{h D_Z}{\varepsilon}. \qquad (2.29)$$

From Eqs. (2.4) and (2.24), the current into the positive electrode can be expressed in terms of the electrode surface charge density as

$$I_C = A\frac{d\rho_S}{dt} = A\frac{dD_N}{dt} \tag{2.30}$$

where the flux density is pointing away from the positive upper electrode so that $D_N = -D_Z$. The total derivative is applicable since we have assumed that ρ_S and D_N are uniform over the electrode surface. If they vary over the electrode surface, then the partial time derivative should be used. Combining Eqs. (2.1), (2.29), and (2.30), we obtain

$$I_C = C\frac{dV_C}{dt} = C\frac{d\left(\frac{-bD_Z}{\varepsilon}\right)}{dt} = A\frac{d(-D_Z)}{dt}. \tag{2.31}$$

Viola! Equating the third and fourth terms, we have an expression for capacitance in terms of the physical parameters of the capacitor as

$$C = \frac{\varepsilon A}{b}. \tag{2.32}$$

This single value, C, is rather deceptive since it "hides" the complex spatial variations of the internal fields just as R doesn't reveal all the geometric information of a resistor. This is especially true when we analyze structures with more complicated field distributions. However, the results are accurate and quite useful in circuits where it is much more convenient to deal with scalar calculations only. In arriving at this result, we have made two important assumptions that will not always be true. First, we assumed that the surface charge density on the electrodes is uniform. In fact, it will be slightly larger at the edges of the parallel plate capacitor and even more complicated in other structures. Secondly, we have assumed that the dielectric acts as an ideal flux guide, hence, no fringing flux. Keep in mind that capacitors with large electrode spacing, b, compared to the shortest electrode dimension and with values of ε_R close to unity there is significant leakage flux. Accurate calculations of leakage flux are difficult and are beyond the scope of this text.

Example 2.7-1.   Calculate the capacitance of a parallel plate capacitor with electrodes that measure 1 cm × 2 cm and with a Teflon dielectric of 0.1 mm thickness. From Appendix D, Teflon has $\varepsilon_R = 2.1$ so that $\varepsilon_{\text{TEFLON}} = 2.1\varepsilon_o = 18.6$ pF/m. The capacitance is calculated by a straightforward application of Eq. (2.32) as

$$C = \frac{\varepsilon A}{b} = \frac{18.6(0.01)(0.02)}{0.0001} = 37.2\,\text{pF}.$$

Example 2.7-2. The capacitor of Example 2.7-1 has a voltage drop of 10 V_{DC}. Calculate the charge on each electrode, the surface charge density, the electric flux density, and the electric field intensity. The electric field for parallel electrodes is given by Eq. (2.27) as $|\mathbf{E}| = V/h = 10/0.0001 = 10^5$ V/m. The related electric flux density is calculated with Eq. (2.28) as $|\mathbf{D}| = \varepsilon|\mathbf{E}| = 18.6 \times 10^{-12}(10^5) = 18.6 \times 10^{-7} = 1.86\,\mu\text{C/m}^2$. $\rho_S = |D_N| = 1.86\,\mu\text{C/m}^2$ from Eq. (2.24). Finally, assuming that the charge density is uniform, the charge on an electrode is calculated as $Q = \rho_S A = 1.86 \times 10^{-6}(0.01)(0.02) = 372$ pC, plus on the positive electrode, minus on the negative one.

2.8   CAPACITANCE: A CLOSER LOOK

Insight into the principles of a capacitor is gained by integrating the third and fourth terms of Eq. (2.31) with respect to time to obtain

$$AD_Z = C\frac{hD_Z}{\varepsilon} = ChE_Z. \qquad (2.33)$$

Recalling that $|D_Z| = \rho_S$ and that the charge is assumed uniformly distributed, we recognize the LHS as the magnitude of the total charge on each of the electrodes. Since the field is uniform, $|hE_Z| = V_C$, the magnitude of the voltage drop across the capacitor. These results lead to an alternate definition of capacitance

$$C = \frac{Q}{V_C} \qquad (2.34)$$

where Q is the magnitude of the charge on each electrode and V_C is the voltage drop across the capacitor. From this expression, we are led to an interpretation of capacitance as a measure of the amount of charge on the electrodes when a 1 V drop is applied. This is a very fundamental definition and from it we can obtain Eq. (2.1) by taking the time derivative and solving for I_C. Rewriting Eq. (2.34) as

$$V_C = \frac{Q}{C} = \frac{h}{\varepsilon}\left(\frac{Q}{A}\right), \qquad (2.35)$$

we gain additional insight into the behavior of capacitors. An increase in area (accompanied by an increase in capacitance) requires an increase in charge on the electrodes in order to maintain a 1 V drop. A decrease in spacing between electrodes (accompanied by an increase in capacitance) requires an increase in charge to maintain the 1 V voltage drop. An increase in permittivity (accompanied by an increase in capacitance) requires an increase in charge to compensate for the increase in bound charge at the dielectric–electrode surface and maintain the 1 V voltage drop.

Equation (2.34) can be modified to handle cases with nonuniform charge distribution and fields. In general, the magnitude of the charge on an electrode is given by Gauss' law as

$$|Q| = \left| \oiint_{\text{ELECTRODE}} \mathbf{D} \cdot \mathbf{ds} \right| = \left| \oiint_{\text{ELECTRODE}} \varepsilon \mathbf{E} \cdot \mathbf{ds} \right| \qquad (2.36)$$

where the surface S encloses the electrode. The magnitude of the voltage drop is

$$|V_C| = \left| \int_{\substack{\text{BETWEEN} \\ \text{ELECTRODES}}} \mathbf{E} \cdot \mathbf{dl} \right| \qquad (2.37)$$

where the integral follows any convenient path between the two electrodes. Since the capacitance is always positive, we use the absolute values to express capacitance as

$$C = \frac{\left| \oiint_{\substack{\text{ELECTRODE} \\ \text{SURFACE}}} \varepsilon \mathbf{E} \cdot \mathbf{ds} \right|}{\left| \int_{\substack{\text{BETWEEN} \\ \text{ELECTRODES}}} \mathbf{E} \cdot \mathbf{dl} \right|}. \qquad (2.38)$$

This is of the same form as Eq. (1.173). The capacitance is expressed solely in terms of the electric fields integrated over the geometry of the capacitor. This equation is deceptively simple since determination of $\mathbf{E}$ is a major task as $\mathbf{E}$ is greatly dependent upon the capacitor geometry as well. This expression may not be especially useful for calculations, but it brings out the importance of the electric field in calculating capacitance. To be exact the surface integral should be integrated over the entire electrode surface. But when the dielectric acts as an ideal flux guide, the contribution of the fields outside of the dielectric will be negligible; the integral can be limited to the surface of the electrode–dielectric interface where the charge is located. This expression is valid whether the charge distribution is uniform or has spatial variation on the electrode.

A somewhat different perspective of capacitance follows from Gauss' law as well: the value of the electric flux emanating from an electrode is equal to the charge upon it. Accordingly,

Eq. (2.34) is modified to

$$C = \frac{\Psi_e}{V_c}. \qquad (2.39)$$

Capacitance is the ratio of the electric flux through the capacitor to the voltage drop across it.

This view doesn't offer any advantage over calculation methods covered earlier in this chapter. But, note that this relationship is analogous to the relationship describing conductance as the ratio of the total current flux through the resistor to the voltage drop across it, i.e., $G = \Psi_I/V_R$. When applied to a capacitor that is divided into incremental regions, series–parallel circuit techniques can be used to calculate capacitance, see Fig. 2.9.

Each incremental flux tube contains a flux $\Delta\Psi_e$. Incremental voltage drops of ΔV are shown for each flux tube. The fields within this region behave as an incremental capacitor in that the flux and equipotentials are mutually perpendicular. Though the flux lines are a continuation of lines from adjacent cells, their behavior is the same as if PEC electrodes with a voltage drop

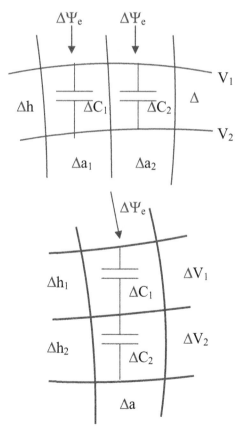

FIGURE 2.9: Incremental capacitors: (a) parallel and (b) series.

ΔV were inserted along the equipotentials of the region. This region forms an incremental capacitor that approximates a parallel plate capacitor with an ideal flux guide. Equation (2.39) can be used to calculate the capacitance as

$$\Delta C = \frac{\Delta \Psi_e}{\Delta V} = \frac{\varepsilon \Delta w \Delta L}{\Delta h}. \tag{2.40}$$

Incremental capacitors in series–parallel combinations fill the region between the capacitor electrodes. This leads to the curvilinear squares method for capacitance calculations. The capacitance per square (for $\Delta L = \Delta h$ and $\Delta w = t$) is given by $C_\square = \varepsilon t$.

There is a remarkable similarity between the fields of resistors and capacitors. Correspondingly, element value calculations for both can be defined in terms of flux concepts as

$$G = \frac{1}{R} = \frac{\Psi_I}{V_R} \quad \text{and} \quad C = \frac{\Psi_e}{V_C}. \tag{2.41}$$

The element values of both elements can be viewed as the flux produced within the element per unit voltage drop across the element. Without a doubt, the nature of the two fluxes is quite different, but the principles of that they obey and the calculations are the same. In fact, comparing the equations for cylindrical resistors and parallel plate capacitors, we observe

$$G = \frac{\sigma A}{L} \quad \text{and} \quad C = \frac{\varepsilon A}{h}. \tag{2.42}$$

This similarity in the form of capacitance and resistance expressions is valid in general as well. For identical configurations, this relationship is often expressed as

$$RC = \left(\frac{\Psi_i}{V_R}\right)^{-1} \frac{\Psi_e}{V_C} = \frac{\Psi_e}{\Psi_i} = \frac{\varepsilon}{\sigma}. \tag{2.43}$$

This suggests that all the methods we learned for resistance calculations can be applied to capacitance as well. As you may have already guessed, inductors will have a similar form too! The reason for identical underlying principles for capacitors and resistors is discussed in the following section.

2.9   LAPLACE'S EQUATION REVISITED

The fields within dielectric materials satisfy Gauss' law in point form as

$$\nabla \cdot \mathbf{D} = \rho_V. \tag{2.44}$$

Substitution of $\mathbf{D} = \varepsilon \mathbf{E}$ and application of the chain rule leads to

$$\nabla \cdot \mathbf{D} = \nabla \cdot (\varepsilon \mathbf{E}) = \nabla \varepsilon \cdot \mathbf{E} + \varepsilon \nabla \cdot \mathbf{E} = \rho_V \tag{2.45}$$

where the "chain rule" of vector calculus (see Appendix A) has been employed. Since most dielectric materials are uniform, their spatial derivatives vanish. We will assume the permittivity is constant so that $\nabla \varepsilon = 0$. In addition, substitution of $\mathbf{E} = -\nabla V$ leads to

$$\nabla \cdot \mathbf{D} = \varepsilon \nabla \cdot \mathbf{E} = -\varepsilon \nabla \cdot \nabla V = -\varepsilon \nabla^2 V = \rho_V \qquad (2.46)$$

where the last two terms on the RHS can be rewritten in the form of Poisson's equation

$$\nabla^2 V = -\frac{\rho_V}{\varepsilon}. \qquad (2.47)$$

The voltage depends upon the charge density. In addition, fixed boundary voltages also affect the solution. This inhomogeneous equation is difficult to solve directly. Fortunately, most dielectrics have no free charge, i.e., $\rho_V = 0$, which leads to Laplace's equation

$$\nabla^2 V = 0. \qquad (2.48)$$

We already solved this same equation in finding the voltage distribution in resistors. In fact, for identical electrode and flux tube geometry, the spatial distribution of voltages is the same for resistors and capacitors.

2.10   ELECTRIC ENERGY STORAGE

The instantaneous power flow into a capacitor is calculated from the circuit concepts by

$$P_C = V_C I_C = C V_C \frac{dV_C}{dt} = C \frac{d}{dt} \left(\frac{V_C^2}{2} \right) \qquad (2.49)$$

where I_C is the current flow in the direction of the voltage drop V_C. The power delivered to the capacitor is positive when V_C^2 is increasing; it is negative when V_C^2 is decreasing; it is zero when the voltage is unchanging. If we think of power as the time derivative of energy, i.e., $P_C = dW_C/dt$, from Eq. (2.49) we obtain

$$W_C = \frac{C V_C^2}{2}. \qquad (2.50)$$

From Eq. (2.50), we can see that no energy is stored when the voltage drop is zero; the sign of energy is always positive regardless of the polarity of the applied voltage. The process can now be described more completely. While the voltage increases, power is delivered to the capacitor as the energy stored in the capacitor is increasing. While the voltage decreases, the capacitor delivers power back to the circuit as the energy stored in the capacitor decreases. While the voltage is constant, no power flow occurs and the energy stored is unchanged.*

---

* This description assumes an ideal dielectric without any losses.

Since the voltage drop is related to the electric field, the energy stored in the capacitor can be expressed in terms of the fields. Equation (2.50) becomes

$$W_C = \frac{CV_C^2}{2} = \frac{1}{2}\left(\frac{\varepsilon \iint\limits_S \mathbf{E} \cdot \mathbf{ds}}{\int\limits_L \mathbf{E} \cdot \mathbf{dl}}\right)\left(\int\limits_L \mathbf{E} \cdot \mathbf{dl}\right)^2$$

$$= \frac{1}{2}\left(\varepsilon \iint\limits_S \mathbf{E} \cdot \mathbf{ds}\right)\left(\int\limits_L \mathbf{E} \cdot \mathbf{dl}\right). \tag{2.51}$$

The electric field occurs twice in this expression, so that energy is proportional in some way to the square of the electric field.

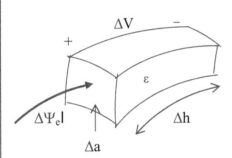

FIGURE 2.10: Incremental capacitor.

An alternate approach provides more insight into this relationship. Consider the incremental capacitor shown in Fig. 2.10. An expression for the incremental energy stored is obtained from Eq. (2.50) as

$$\Delta W_C = \frac{\Delta C(\Delta V_C)^2}{2} = \frac{\varepsilon \Delta a}{2\Delta l}(|\mathbf{E}|\Delta l)^2$$

$$= \frac{\varepsilon|\mathbf{E}|^2}{2}\Delta\forall \tag{2.52}$$

where $\Delta\forall$ is the volume of the incremental capacitor. The quantity which multiplies $\Delta\forall$ has units of energy/m³ and represents the *stored electric energy density* or just *electric energy density* and is represented as

$$w_e = \frac{\Delta W_C}{\Delta\forall} = \frac{\varepsilon|\mathbf{E}|^2}{2} = \frac{\varepsilon\mathbf{E} \cdot \mathbf{E}}{2}$$

$$= \frac{\mathbf{E} \cdot \mathbf{D}}{2} \quad [\text{J/m}^3] \tag{2.53}$$

with units of Joules/cubic meter. The subscript "e" is to signify electric energy density. It turns out that this is a correct representation for electric energy density anywhere an electric field exists. The total energy stored in a volume can be written as

$$W_e = \sum_{\text{VOLUME}} \Delta W_e = \sum_{i=1}^{N} w_{ei}\Delta v_i = \sum_{i=1}^{N} \frac{\varepsilon_i|\mathbf{E}_i|^2}{2}\Delta v_i \tag{2.54}$$

where i represents the ith volume. Note that ε and $\mathbf{E}$ can vary from throughout the region; their values at the center of each cell are used to compute the energy within that cell. This form is especially good when incremental forms of ε and $\mathbf{E}$ are available. When they vary continuously, an integral form is more useful

$$W_e = \iiint_\forall \frac{\varepsilon|\mathbf{E}|^2}{2} dv. \tag{2.55}$$

Since the main function of capacitors is to store energy, a definition of capacitance based upon energy stored is considered to be more fundamental than the more familiar circuit definition of Eq. (2.1). This can be achieved by combining Eqs. (2.51) and (2.55) as

$$C = \frac{2W_e}{V_C^2} = \frac{2 \iiint_\forall \frac{\varepsilon|\mathbf{E}|^2}{2} dv}{\left(\int_L \mathbf{E} \cdot \mathbf{dl}\right)^2} = \frac{\iiint_\forall \varepsilon|\mathbf{E}|^2 dv}{\left(\int_L \mathbf{E} \cdot \mathbf{dl}\right)^2}. \tag{2.56}$$

Optimization methods for calculating capacitance utilize the fundamental principle that the correct solution is that value of $\mathbf{E}$ which minimizes the energy stored for a given voltage drop—that's nature's way. This form is preferred for such calculations. But we will defer these more advanced methods to more advanced courses.

Finally, the concept of energy density is often useful when modeling high frequency and microwave circuits. Those regions in which there is a high electric energy density are considered to be capacitive in nature and frequently are modeled by lumped capacitive elements. Equation (2.56) is used to calculate this equivalent capacitance.

2.11   CAPACITANCE CALCULATIONS

A variety of concepts and solution methods for calculating capacitance are presented in the previous sections. In this section, they are illustrated by several applications.

Example 2.11-1. Calculate the capacitance of a 1 m length section of RG58 coaxial cable such as you use in circuits laboratory to connect a circuit to a generator or a meter. From manufacturer's catalogs or reference data handbooks, we find that RG58 cable has a center conductor of about 0.4 mm radius, a polyethylene dielectric of 1.5 mm radius surrounded by an outer copper braid*, see Fig. 2.11.

---

* There are several versions of the RG58/X transmission line where the X indicates the version. We will use specifications that apply nominally to all of them.

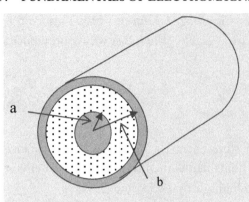

FIGURE 2.11: Coaxial cable.

From Appendix D, we find $\varepsilon_{\text{TEFLON}} = 2.26\varepsilon_o = 20.0$ pF/m. Since the electric fields of the coaxial capacitor and the coaxial resistor both satisfy Laplace's equation and the same boundary conditions, they are the same as expressed in Example 1.29-1 as

$$\mathbf{E} = \frac{V_O}{\ln(b/a)\,\rho}\mathbf{a}_\rho = \frac{V_O}{\ln(0.0015/0.0004)\,\rho}\mathbf{a}_\rho$$

$$= \frac{0.757V_O}{\rho}\mathbf{a}_\rho$$

where the voltage drop from the inner to the outer conductors is V_O. Direct calculation of capacitance requires the calculation of charge on the electrodes. The charge is related to the surface charge density which is equal to the electric flux density at the surface of the electrode as

$$\rho_S|_{\rho=.0004} = |\mathbf{D}|_{\rho=.0004} = \varepsilon\,|\mathbf{E}|_{\rho=.0004}$$

$$= \frac{0.757\varepsilon V_O}{0.0004}\mathbf{a}_\rho = 1891\varepsilon V_O\mathbf{a}_\rho$$

and

$$Q_a/m = \iint_S \rho_S|_{\rho=0.0004}\,da = \int_{z=0}^{1}\int_{\phi=0}^{2\pi} 1891\varepsilon V_O\,(0.0004d\phi dz)$$

$$= 4.75\varepsilon V_O \text{ C/m}.$$

From Eq. (2.29), the capacitance is given as $C/m = \frac{Q_a/m}{V_O} = 4.75(20.0) = 95.0\,\text{pF/m}$. The usual parameter to describe the capacitance of a coaxial cable is as capacitance/meter; that's just what we have calculated.

Alternatively, we could have calculated the surface charge density on the outer electrode as $\rho_S|_{\rho=0.0015} = -\frac{0.757\varepsilon V_O}{0.0015} = -504.7\varepsilon V_O$; integration over the outer electrode gives $Q_a/m = -4.75\varepsilon V_O$ C/m. The charge density is less since the surface area of the outer electrode is greater, but the total charge is the same (as it must be) .

As a further alternative, we could have calculated the capacitance directly by using the form of the resistance expression of Example 1.29-1 and Eq. (1.43) as

$$C = \frac{2\pi\varepsilon L}{\ln(b/a)} = \frac{2\pi(20.0)(1)}{\ln(0.0015/0.0004)} = 95.0 \text{ pF/m}.$$

Finally, let's use the energy definition of capacitance, Eq. (2.56), as

$$C = \frac{\iiint_{\forall} \varepsilon|\mathbf{E}|^2 dv}{\left(\int_L \mathbf{E} \cdot \mathbf{dl}\right)^2} = \frac{\varepsilon \int_{z=0}^{L} \int_{\phi=0}^{2\pi} \int_{\rho=a}^{b} \left(\frac{V_0}{\ln(b/a)\,\rho}\right)^2 \rho\,d\rho\,d\phi\,dz}{\left(\int_{\rho=a}^{b} \frac{V_0}{\ln(b/a)\,\rho}\,d\rho\right)^2}$$

$$= \frac{\dfrac{2\pi\varepsilon L V_0^2 \ln(b/a)}{(\ln(b/a))^2}}{\dfrac{V_0^2 (\ln(b/a))^2}{(\ln(b/a))^2}} = \frac{2\pi\varepsilon L}{\ln(b/a)},$$

the same form we obtained before.

Example 2.11-2. Calculate the capacitance/meter of the square coaxial transmission line shown in Fig. 2.12 by the method of curvilinear squares. The space between the PEC electrodes is filled with aluminum oxide. From Appendix D, we find that $\varepsilon_R = 8.8$. Using the analogy between conductance and capacitance expressed by Eq. (2.41), we can express the capacitance of each curvilinear square a $C_{\text{SQUARE}} = \varepsilon t$ so that the capacitance/meter length for each square is given by $C/m = \varepsilon$. Moreover due to the symmetry of the structure, we need only to sketch curvilinear squares for one-eighth of the coaxial cross section and multiply the result by eight (capacitors in parallel add) to obtain the total capacitance. The electric flux is di-

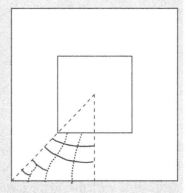

FIGURE 2.12: Capacitance by curvilinear squares.

rected from the inner square electrode to the outer square electrode. There are three flux tubes, the rightmost two tubes have three capacitors in series; the leftmost tube is more complicated. From the squares shown in Fig. 2.12, we find that the capacitance is given by

$$C/m = 8\left(\frac{1}{3} + \frac{1}{3} + \frac{5}{13}\right)\varepsilon = 7.52\varepsilon_R\varepsilon_o$$

$$= 7.52(8.8)(8.854 \times 10^{-12}) = 585.5 \text{ pF/m}.$$

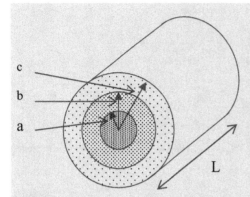

c

b

a

L

FIGURE 2.13: Inhomogeneous coaxial capacitor.

Example 2.11-3. Calculate the capacitance of the inhomogeneous coaxial capacitor shown in Fig. 2.13. The permittivity between the two PEC electrodes varies as 1 cm $\leq \rho \leq 3$ cm, $\varepsilon_R = 2$ and 3 cm $\leq \rho \leq 5$ cm, $\varepsilon_R = 1$. The boundary between the two dielectrics is located at $\rho_{\text{DIELECTRIC}} = b = 3$ cm. The PEC electrodes are located at $\rho_{\text{INNER}} = a = 1$ cm and $\rho_{\text{OUTER}} = c = 5$ cm, respectively. The capacitor is $L = 5$ cm long.

Due to the symmetry of the capacitor, the dielectric–dielectric interface coincides with an equipotential surface. Consequently, the two dielectrics can be treated as two capacitors in series; the capacitance of each is calculated according to Example 2.11-1 as $C = \frac{2\pi \varepsilon L}{\ln(b/a)}$ so that $C_{\text{INNER}} = \frac{2\pi (2\varepsilon_o)(0.05)}{\ln(0.03/0.01)} = 5.06$ pF and $C_{\text{OUTER}} = 5.45$ pF. The combined capacitance is

$$C_{\text{COAX}} = \left(\frac{1}{C_{\text{INNER}}^{-1} + C_{\text{OUTER}}^{-1}} \right)^{-1} = 2.62 \text{ pF}.$$

Example 2.11-4. Calculate the capacitance/meter of the coaxial transmission line of Example 2.11-2 by the spreadsheet method. Symmetry can be used advantageously in numeric methods as well; due to the rectangular nature of the spreadsheet, it is convenient to use one-fourth of the structure, see Fig. 2.14. Each of the nodal voltages satisfies Eq. (2.58). The capacitance of the quadrant is given by $C_{\text{QUAD}} = Q_C/V_C = \Psi_e/1 = \Psi_e$ where the total flux to/from an electrode is calculated as with resistors.

The electric flux in the vicinity of nodes 1, 2, 11, and 16; the calculation is of the form of edge calculations given by Eq. (1.154) since the flux at these nodes is all directed from the inner electrode to the outer. Determination of capacitance is based upon the calculation of the electric flux at either electrode in the manner analogous to the calculation of current flux of Eq. (1.155). Due to symmetry, the electric flux on the left and bottom surfaces of the electrodes are the same, allowing simplification of the calculation; the calculation equation is shown just below the spreadsheet calculations. The results are comparable to Example 2.11-2.

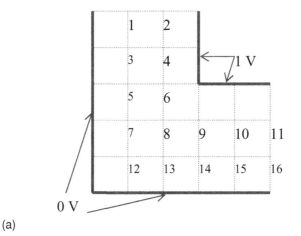

(a)

	A	B	C	D	E	F	G
1	0	0.310	0.641	1			
2	0	0.299	0.626	1			
3	0	0.262	0.565	1	1	1	
4	0	0.181	0.373	0.565	0.626	0.641	
5	0	0.091	0.181	0.262	0.299	0.310	
6	0	0	0	0	0	0	
7	C/m=4*(2*sum(B5:E5)+F5)*8.8*8.854=					615.85	pF/m

(b)

FIGURE 2.14: Capacitance by spreadsheet methods: (a) geometry and (b) calculations.

2.12 FORCES AND VIRTUAL WORK

"Opposite charges attract, like charges repel" is well known to high school and college physics students. This fundamental principle is at work in capacitors, setting up forces between the two electrodes. The details of this force and its magnitude are the subject of this section.

The energy stored in a parallel electrode capacitor is calculated from Eq. (2.55). When ε and $\mathbf{E}$ are constant throughout the capacitor, this becomes

$$W_e = \iiint_\forall \frac{\varepsilon |\mathbf{E}|^2}{2} dv = \frac{\varepsilon (V_C/h)^2}{2} Ah$$

$$= \frac{\varepsilon V_C^2 A}{2h} = \frac{C V_C^2}{2}. \tag{2.57}$$

Alternatively, we can evaluate the integral using $|\mathbf{D}| = Q/A$ to obtain

$$W_e = \iiint_\forall \frac{\varepsilon|\mathbf{E}|^2}{2} dv = \frac{1}{\varepsilon}\left(\frac{Q}{A}\right)^2 Ah$$

$$= \frac{Q^2 h}{2\varepsilon A} = \frac{Q^2}{2C}. \tag{2.58}$$

The stored electric energy is expressed in terms of either the voltage drop across the capacitor or the charge on its electrodes.

To determine the force between the two electrodes (call this the electrical force), we perform a virtual experiment in which one electrode is moved an incremental distance. As the electrode is moved, the capacitance, the ratio of Q/V_C, and the stored electric energy change. We will consider two separate cases: the charge remains constant (the electrodes are open-circuited so that no charge can be conducted away) and the voltage drop remains constant (the electrodes are connected to a fixed power supply).

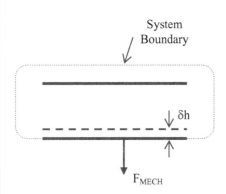

System Boundary

$\downarrow \delta h$

F_{MECH}

FIGURE 2.15: Virtual work system model for constant charge.

For the case of constant charge, we use Eq. (2.58) for calculations. The mechanical force will add energy to the system that is converted to stored electric energy within the capacitor. But, since there are no losses in the system, the total energy of the system must be conserved, see Fig. 2.15. The virtual change in stored electric energy of the capacitor is expressed as

$$\delta W_e = \delta\left(\frac{Q^2}{2C}\right) = \delta\left(\frac{Q^2 h}{2\varepsilon A}\right) = \frac{Q^2 \delta h}{2\varepsilon A}. \tag{2.59}$$

An increase in spacing will increase the stored electrical energy within the capacitor.

Mechanical energy is added to the system by the external mechanical force, F_{MECH}, as it displaces an electrode a virtual distance δh; the virtual mechanical energy added is calculated from elementary mechanics as

$$\delta W_{\text{MECH}} = F_{\text{MECH}}\delta h. \tag{2.60}$$

Since system energy is conserved, the change in stored electrical energy must equal the mechanical energy supplied, i.e., $\delta W_{\text{MECH}} = \delta W_e$. The mechanical energy is calculated by equating Eqs. (2.59) and (2.60) as

$$F_{\text{MECH}} = \frac{Q^2}{2\varepsilon A}. \tag{2.61}$$

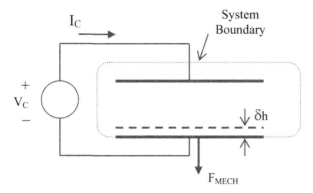

FIGURE 2.16: Virtual work system model for constant voltage.

Since the RHS is positive, the mechanical force is positive in the direction of increasing h. The mechanical force must be exerted because there is an equal and oppositely directed electrical force that tends to pull the electrodes together. This force is expressed by

$$F_e = -F_{\text{MECH}} = -\frac{Q^2}{2\varepsilon A}. \tag{2.62}$$

For the case of constant voltage, the procedure is similar, but Eq. (2.57) is used to represent the energy stored in the capacitor and there is the possibility of electrical energy flow into the system from the power supply, see Fig. 2.16. Following the procedure for constant charge, we obtain

$$\delta W_e = \delta \frac{V_C^2 \varepsilon A}{2h} = \frac{V_C^2 \varepsilon A}{2} \delta\left(\frac{1}{h}\right) = -\frac{V_C^2 \varepsilon A}{2h^2} \delta h \tag{2.63}$$

and

$$\delta W_{\text{MECH}} = F_{\text{MECH}} \delta h. \tag{2.64}$$

The voltage on the capacitor is kept constant via its connection with a voltage source. But, as the electrodes are separated, the accompanying reduction in capacitance causes charges to flow from the capacitor to the voltage source. This charge flow returns some of the capacitor's energy to the voltage source since the charge flows into the positive terminal of the source. The energy supplied to the system by the voltage source is

$$\delta W_S = V_C I_C \delta t = V_C \delta Q = V_C (V_C \delta C)$$
$$= V_C^2 \delta C = V_C^2 \left(-\frac{\varepsilon A}{h^2} \delta h\right) = -\frac{V_C^2 \varepsilon A}{h^2} \delta h. \tag{2.65}$$

The negative sign confirms that the system returns energy to the power supply. Since energy is conserved, the change in system energy must equal the energy supplied by the mechanical force and the electrical source as

$$\delta W_e = \delta W_{\text{MECH}} + \delta W_S \tag{2.66}$$

which is solved to give the mechanical force required to pull apart the electrodes as

$$F_{\text{MECH}} = -\frac{V_C^2 \varepsilon A}{2h^2} + \frac{V_C^2 \varepsilon A}{h^2} = \frac{V_C^2 \varepsilon A}{2h^2} = \frac{Q^2}{2\varepsilon A}. \tag{2.67}$$

The mechanical force reduces the stored energy within the system as charges leave the capacitor which in turn provides energy to the power supply. As in the constant charge case, the mechanical force is exerted against the electrical force which tends to pull the electrodes together and

$$F_e = -F_{\text{MECH}} = -\frac{V_C^2 \varepsilon A}{2h^2} = -\frac{Q^2}{2\varepsilon A}. \tag{2.68}$$

It is reassuring that Eq. (2.68) agrees with Eq. (2.60) since the force should have no dependence upon which "virtual experiment" we base our calculations. The electrical force depends upon the square of the voltage or the charge and tends to pull the electrodes together. This agrees with our intuition that the opposite charges on the two electrodes tend to attract each other. Actually the forces are between the charges residing on the electrodes, not the electrodes themselves. The fields due to the charges on one electrode exert forces on the charges of the other electrode. These charges are usually electrons that are relatively mobile. However, in most applications, the atoms of the metallic electrodes exert a stronger force on the electrons than that of the external field and the electrons are held captive by the atoms of the electrodes. Consequently, the forces on the electrons are transferred to the electrode via the atoms.

The electric force on an electrode as given by Eqs. (2.62) and (2.68) can be generalized as

$$F_e = -\frac{V_C^2 \varepsilon A}{2h^2} = -\frac{\varepsilon |\mathbf{E}|^2 A}{2} = -w_e A, \tag{2.69}$$

where w_e is the electric energy density within the dielectric adjacent to the PEC electrode. This is commonly expressed as pressure as

$$p_e = \frac{F_e}{A} = -\frac{\varepsilon |\mathbf{E}|^2}{2} = -w_e. \tag{2.70}$$

The pressure on an electrode is equal to the energy density at the electrode and tends to pull the electrodes together. There is no tangential component of force or pressure at a dielectric–electrode interface.

The virtual work method reveals the forces at dielectric–dielectric interfaces as well. For case where the electric field is tangential to the interface, see Fig. 2.17(a), the electric field is the same on both sides of the interface, $E_{T1} = E_{T2} = E_T$, and is unchanged by a shift in the interface. A virtual displacement of the boundary into region 1 results in a change energy given by

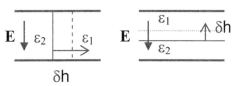

FIGURE 2.17: Virtual displacement of dielectric boundaries: (a) tangential field and (b) normal field.

$$\delta W_e = \left[\frac{\varepsilon_2 |\mathbf{E}_T|^2}{2} - \frac{\varepsilon_1 |\mathbf{E}_T|^2}{2}\right]\delta h\, A$$

$$= (\varepsilon_2 - \varepsilon_1)\frac{|\mathbf{E}_T|^2 \delta h\, A}{2}$$

$$= [w_{T2} - w_{T1}]\delta h\, A. \tag{2.71}$$

Since $\mathbf{E}$ is proportional to V, the same principles used in the constant voltage case as expressed in Eq. (2.68) leads to

$$p_{Tin} = \frac{F_e}{A} = \frac{\delta W_e}{A\delta h} = \frac{(\varepsilon_2 - \varepsilon_1)|\mathbf{E}_T|^2}{2}$$

$$= w_{T2} - w_{T1} \tag{2.72}$$

pointing into region 2 where w_{Ti} represents the energy density of the tangential component of the field in the ith region. The **electric pressure tends to push the boundary toward the dielectric having the smaller permittivity**.

When the electric field is normal to the boundary, see Fig. 2.17(b), $D_{N1} = D_{N2} = |\mathbf{D}_N|$, and the change in electric energy by a virtual displacement into region 1 is

$$\delta W_e = \left[\frac{\varepsilon_2 |\mathbf{E}_{N2}|^2}{2} - \frac{\varepsilon_1 |\mathbf{E}_{N1}|^2}{2}\right]\delta h\, A$$

$$= \left[\frac{|\mathbf{D}_N|^2}{2\varepsilon_2} - \frac{|\mathbf{D}_N|^2}{2\varepsilon_1}\right]\delta h\, A$$

$$= \left(\frac{1}{\varepsilon_2} - \frac{1}{\varepsilon_1}\right)\frac{|\mathbf{D}_N|^2 \delta h\, A}{2}$$

$$= [w_{N2} - w_{N1}]\delta h\, A. \tag{2.73}$$

Since $D_{N1} = D_{N2} = |\mathbf{D}|$ is unaffected by a shift in the boundary, this case compares with the constant charge case as expressed by Eq. (2.64) to give

$$p_{Nin} = \frac{F_e}{A} = -\frac{\delta W_e}{A\delta h} = \left(\frac{1}{\varepsilon_1} - \frac{1}{\varepsilon_2}\right)\frac{|D_N|^2}{2}$$

$$= w_{N1} - w_{N2} \tag{2.74}$$

where w_{Ni} represents the energy density of the field normal to the interface. As before, the pressure tends to push the interface into the region with the lesser dielectric.

Similarly, torques are generated by the forces on electrodes or dielectric interfaces. In addition, this technique will be very helpful in analyzing magnetic devices as well.

Example 2.12-1.   Calculate the pressure and the force on the electrodes of a parallel plate capacitor of Example 2.7-2 when 10 V is applied. Since the electric field is constant throughout the capacitor as $|\mathbf{E}| = 10^5$ V/m, then the energy density and pressure are given as $p_e = w_e = \varepsilon|\mathbf{E}|^2/2 = 18.6 \times 10^{-12}(10^5)^2/2 = 0.093$ N/m². The force is $F_e = p_e A = 0.093(0.0002) = 0.0000186$ N. Note that 1 N is the force exerted by gravity on about 0.1 kg mass—about 4 oz. This is a rather small force. A 100 V source applied would still produce a very small force. Larger voltage drops would produce $|\mathbf{E}| > 10^6$ V/m which is getting close to the critical field strength of air with the possibility of causing an arc at the edge of the electrodes.

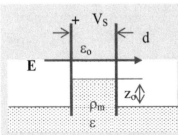

FIGURE 2.18: Lifting of a fluid between two electrodes.

Example 2.12-2.   Two electrodes separated by distance d are immersed in a dielectric fluid with permittivity ε as shown in Fig. 2.18. A DC voltage source V_S is applied across the electrodes. The fluid has a mass density of ρ_m. Calculate the height z which the fluid between the electrodes is lifted above the level of the surrounding reservoir.

The electric field between the two electrodes is parallel to the interface between the dielectric fluid and the air. We calculate the interface pressure that tends to lift the fluid as $p_{Tin} = w_{T2} - w_{T1} = \frac{(\varepsilon - \varepsilon_o)}{2}\left(\frac{V_S}{d}\right)^2$. At the same time, gravity is exerting a downward pressure of $p_{GRAVITY} = \rho_m g z_o$, where z_o is the height of the fluid between the electrodes above the reservoir level. At equilibrium, these pressures must be equal which leads to the height of the fluid between the two electrodes as $z_o = \frac{(\varepsilon - \varepsilon_o)}{2\rho_m g}\left(\frac{V_S}{d}\right)^2$. For a spacing of 1 cm, a source of 100 V, and the fluid of distilled water ($\rho_m = 1$ and $\varepsilon = 80\varepsilon_o$), the height is $z_o = \frac{(80\varepsilon_o - \varepsilon_o)}{2(9.8)}\left(\frac{100}{0.01}\right)^2 = 3.6$ mm. This is hardly visible, but with a voltage of 1000 V this becomes

36 cm, a quite significant lifting effect. Applications of this principle in the micro-gravity conditions of earth's orbit increase this effect significantly.

2.13   FLUX DENSITY VIA GAUSS' LAW

There are a significant number of applications in which the dielectric flux guide is not nearly ideal. The electric flux emanating from charges is not confined, but reaches everywhere. In order to handle these situations, we must take a different approach to calculating the electric flux density.

Gauss' law can be used to find the electric flux density for several common charge distributions. A point charge, as shown in Fig. 2.19, is particularly useful. The electric flux density of a point charge has spherical symmetry, i.e., $\mathbf{D}(r, \theta_1, \phi_1) = \mathbf{D}(r, \theta_2, \phi_2)$, so that it varies only with r, i.e., $\mathbf{D} = \mathbf{D}(r)$. From Eq. (2.6), we observe that there is a net outflow of flux lines from the surface S only when it contains charge. As the surface S is made smaller and smaller, the only flux lines out of S are those which begin or end within the region. Since the small sphere encloses only the small point charge, the flux lines emanate outward from charge itself. Therefore, flux lines begin and end on a point charge and are radially directed outward from it, i.e., $\mathbf{D} = D_r(r)\mathbf{a}_r$. The directed surface element for the spherical surface S is given by $\mathbf{ds} = r^2 \sin\theta d\theta d\phi \mathbf{a}_r$. Combining these details, we obtain

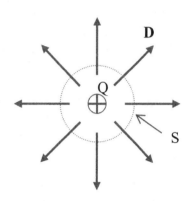

FIGURE 2.19:   Electric flux of a point charge.

$$\Psi_e = \int_{\phi=0}^{2\pi} \int_{\theta=0}^{\pi} D_r(r)\mathbf{a}_r \cdot \mathbf{a}_r r^2 \sin\theta d\theta d\phi$$

$$= D_r(r)r^2 4\pi = Q_{\mathrm{PT}}. \tag{2.75}$$

Therefore, $D_r(r) = Q_{\mathrm{PT}}/4\pi r^2$, so that

$$\mathbf{D} = \frac{Q_{\mathrm{PT}}\mathbf{a}_r}{4\pi r^2} \ \mathrm{C/m^2}. \tag{2.76}$$

The flux density of a point charge points directly away from a point charge and shows no dependence upon the material surrounding the charge. But of more fundamental importance, it shows r^{-2} functional dependence. With this form, the dot product of the flux density and the spherical differential surface element is given by $(Q/4\pi) \sin\theta d\theta d\phi$ which is independent of r. Consequently, the flux crossing all spherical surfaces concentric with the point charge, regardless of radius, will be the same, as required by Gauss' law, since all of the concentric spheres enclose the same point charge. With these facts, we are prepared to find the electric flux density for several other charge distributions.

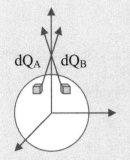

FIGURE 2.20: Radially dependent, spherically symmetric charge distribution.

Example 2.13-1. Consider the radially dependent, spherically symmetric charge distribution shown in Fig. 2.20.

Two differential volumes of charge, dQ_A and dQ_B, are located symmetrically with respect to the z-axis. Due to their differentially small volume, they act as point charges for which Eq. (2.63) describes the electric flux density. The two volumes contribute equally to the electric flux density at point P where their combined result is entirely radial; all other components cancel. All of the charges within the distribution can be matched in similar pairs which likewise contribute only radial components of electric flux density at point P. Due to angular symmetry of the charge distribution, the flux density shows no angular dependence. Combining these details, we obtain an expression for flux which is similar to Eq. (2.75) as

$$\Psi_e = \int\limits_{\phi=0}^{2\pi} \int\limits_{\theta=0}^{\pi} D_r(r)\mathbf{a}_r \cdot \mathbf{a}_r r^2 \sin\theta d\theta d\phi$$

$$= D_r(r)r^2 4\pi = Q_{\text{ENC}}.$$

For $r \geq a$, Q_{ENC} is equal to the total charge within the distribution as

$$Q_{\text{ENC}} = \iiint\limits_{V(r\leq a)} \rho_V dv = 4\pi \int\limits_{r'=0}^{a} \rho_V r'^2 dr' \text{ and } \mathbf{D} = \frac{4\pi \int\limits_{r'=0}^{a} \rho_V r'^2 dr'}{4\pi r^2}\mathbf{a}_r = \frac{\int\limits_{r'=0}^{a} \rho_V r'^2 dr'}{r^2}\mathbf{a}_r.$$

For $r < a$, the charge enclosed depends upon radius as

$$Q_{\text{ENC}}(r) = \iiint\limits_{V(r'\leq r)} \rho_V dv' = 4\pi \int\limits_{r'=0}^{r} \rho_V r'^2 dr' \text{ and } \mathbf{D} = \frac{\int\limits_{r'=0}^{r} \rho_V r'^2 dr}{r^2}\mathbf{a}_r.$$

This example shows that the flux density at radius r for spherically symmetric charge densities is the same as that of a point charge at the center of the distribution and equal to the charge contained within the radius r. Any other charges with the same spherical distribution but at radii greater than that of point P produce fields at point P that cancel out each other's effects.

The direct calculation of $\mathbf{D}$ from Gauss' law by this method is possible for only three common geometries: (1) spherical geometry with variations of charge density in the radial direction only, (2) infinite cylindrical geometry with variation of charge density in the radial direction only, and (3) infinite planar geometry with variations of charge density only in the direction perpendicular to the infinite plane. However, these cases are worth study since they provide prototypical behavior for spherical, cylindrical, and planar configurations that we can compare with similar geometries. Moreover, many actual charge distributions can be approximated by one of these three special cases.

Two conditions must be met to use this method for calculating $\mathbf{D}$. Firstly, the charge distribution must produce a single component of $\mathbf{D}$ that is aligned with a coordinate direction of Cartesian, cylindrical, or spherical coordinate system. Typically, this will be a radial component in the spherical and cylindrical coordinate systems or one of the Cartesian coordinates. Symmetry is essential to this constraint. Using the $\mathbf{D}$ flux of symmetrically located differential charges, we can determine which components of $\mathbf{D}$ exist for given distribution. Secondly, a closed surface must be chosen on which the $\mathbf{D}$ is either perpendicular or parallel to the surface. For the perpendicular case, $\mathbf{D} \cdot \mathbf{ds} = D_n da$; for the parallel case, $\mathbf{D} \cdot \mathbf{ds} = 0$. Moreover, $\mathbf{D}$ should be constant over the surface so that $\int D_N da$ becomes $D_N A$. Under these conditions, Gauss' law becomes $\int \mathbf{D} \cdot \mathbf{ds} = D_N A = Q$ from which D_N can be calculated.

The calculation of $\mathbf{D}$ by this method generally requires a homogeneous media, i.e., there are no variations in permittivity. However, the method is valid in the special case in which permittivity variations occur only in the direction of $\mathbf{D}$. The calculation of $\mathbf{E}$ follows directly from these results via $\mathbf{E} = \mathbf{D}/\varepsilon$.

Example 2.13-2. Calculate $\mathbf{D}$ for an infinitely long filament of uniformly distributed charge, ρ_L. As shown in Fig. 2.21(a), the fields are expected to be the same for all angular positions around the filament of charge. When the filament is aligned with the z-axis, this symmetry condition is manifest as no variations of the fields with ϕ. Regardless of the location of the observation point relative to the z-axis, there is an infinite amount of charge extending to infinity in both the $+z$ and $-z$ directions. This means that the observer "sees" just as much charge in the $+z$ direction as in the $-z$ direction resulting in the same fields regardless of z-location; there is no variation of the field with z. Therefore, the field varies only with ρ, the radial, perpendicular distance from the z-axis, see Fig. 2.21(b). Two differential elements of charge, dQ_A and dQ_B, are symmetrically located with respect to the z-location of the field point. All of the charge of the filament can be represented by two similar symmetric charges. They contribute equally to the flux at the field point, but their z-components are oppositely directed and cancel, leaving only a ρ-component of the field. Thus far we have used differential elements of charge and the symmetry of the problem to show that the

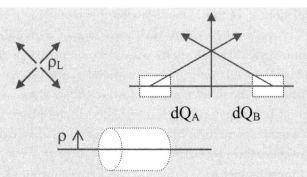

FIGURE 2.21: Infinite filament of uniform charge: (a) end view, (b) side view, and (c) Gaussian surface.

flux density has the form $\mathbf{D} = D_\rho(\rho)\mathbf{a}_\rho$. Now the strategy is to choose a Gaussian surface on which $\mathbf{D} \cdot \mathbf{ds}$ is constant or zero. Note that $\mathbf{D}$ is perpendicular to and constant on the cylindrical portion of the surface that is centered on the z-axis. Furthermore, $\mathbf{D} \cdot \mathbf{ds} = 0$ on the ends of the cylinder. Consequently, a closed cylinder is a good choice as the Gaussian surface, see Fig. 2.21(c). This choice leads to

$$Q_{\text{ENC}} = \rho_L L = \oiint_S \mathbf{D} \cdot \mathbf{ds} = \iint_{\text{CYLINDER}} D_r \rho d\phi dz + \iint_{\text{ENDS}} 0 d\rho dz$$

$$= \int_{z=0}^{L} \int_{\phi=0}^{2\pi} D_r \rho d\phi dz = 2\pi L D_\rho \rho,$$

which leads to $\mathbf{D} = D_\rho(\rho)\mathbf{a}_\rho = \frac{\rho_L}{2\pi\rho}\mathbf{a}_\rho$. The electric flux points radially outward from the uniform filament of charge. Its magnitude varies inversely with ρ. This functional form follows directly from Gauss' law. The surface area of a cylinder varies proportionately with ρ while $\mathbf{D}$ varies inversely with ρ so that their product is constant. It follows that the same total flux passes through every concentric cylindrical surface.

Example 2.13-3.   A uniform volume charge density, ρ_V, of infinite extent in the $x - y$ directions occupies the region $|z| \leq 1$. Calculate $\mathbf{D}$ everywhere. As before, symmetry is used to determine the direction of the field. In this case, use four, equal, differential charges symmetrically located with respect to the z-axis and lying in the same z-plane, dQ_A, dQ_B, dQ_C, and dQ_D, see Fig. 2.22. It is obvious that the x- and y-components of the electric flux cancel each other, but the z-components are equal and add together. Moreover, no matter the (x, y) location of the field point, there is an infinite amount of charge extending in

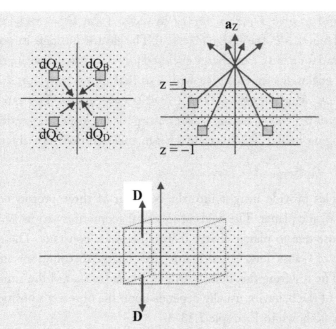

FIGURE 2.22: Planar charge distribution: (a) top view, (b) side view, and (c) Gaussian surface.

the $\pm x$ and $\pm y$ directions. Due to the infinite extent of charge, there are no variations of the electric flux with respect to x and y. Based upon these considerations of charge distribution and symmetry, we deduce that the flux density is of the form $\mathbf{D} = D_Z(z)\mathbf{a}_Z$. The choice of a Gaussian surface is aided by noting that $\mathbf{D} \cdot \mathbf{ds} = 0$ on surfaces described by $x = x_o$ and $y = y_o$. Consequently, a rectangular box with surfaces perpendicular to the Cartesian unit vectors is a good choice since the differential flux $D_Z(z_o)dxdy$ is constant on $z = z_o$ surfaces so that the integral of the flux density becomes $D_Z(z_o)A$. In addition, choice as a Gaussian surface, see Fig. 2.21(c). A final simplification, due to the symmetry of the charge distribution with regard to the $z = 0$ plane, the flux at $z = z_o > 0$ is directed in the $\mathbf{a}_Z$ direction while the flux for $z = -z_o < 0$ is equal, but directed in the $-\mathbf{a}_Z$ direction. First, we calculate the charge contained within the region $|z| \leq 1$ which is given as

$$Q_{\text{ENC}} = \iiint\limits_{\forall} \rho_V dv = \rho_V 2zA$$

$$= \oint\limits_{S} \mathbf{D} \cdot \mathbf{ds} = \iint\limits_{\text{TOP}} D_Z dxdy + \iint\limits_{\text{BOTTOM}} D_Z dxdy = 2AD_Z,$$

which can be solved to give $D_Z(z) = z\rho_V$ or in vector form $\mathbf{D} = \pm D_Z(z)\mathbf{a}_Z = \pm z\rho_V\mathbf{a}_Z$. The $+$ sign is valid for $|z| > 0$, the $-$ sign for $z < 0$. The electric flux density points upward for $z > 0$ and downward for $z < 0$. The charge enclosed for $|z| < 1$ varies with z. For $|z| > 1$, the amount of charge enclosed is unchanging with z so that $Q_{\text{ENC}} = \iiint_V \rho_V dv = \rho_V 2(1)A = \oiint_S \mathbf{D} \cdot \mathbf{ds} = 2AD_Z$ which leads to $\mathbf{D} = \pm D_Z(z)\mathbf{a}_Z = \pm \rho_V\mathbf{a}_Z$. The flux increases as z increases for value of $0 \leq |z| \leq 1$ since greater charge is enclosed for greater $|z|$. However the flux is unchanging for $|z| > 1$ since the charge enclosed does not change for z in this range.

These examples provide insight into the behavior of three prototypical geometries—spherical, cylindrical, and planar. The fields of spherical geometries vary as $1/r^2$ and cylindrical as $1/\rho$ whereas those due to planar charge distributions are constant. These behaviors are a direct consequence of Gauss' law. Often many "real" charge distributions are close to one of these geometries. The resulting flux density will be similar to that of the similar prototype. In addition, the form of the behavior usually depends upon the observer's position relative to the charge distribution as shown in Example 2.13-4.

Example 2.13-4. Consider a long, thin cylindrical structure of length 100 m and diameter of 1 m with a uniform surface charge distribution, ρ_S, over its entire surface. Determine the form of the electric flux density in the mid-plane of the cylinder. For an observer very near the surface, say within 1 mm, the surface "looks" planar so the flux density is nearly constant and is of the form $|\mathbf{D}| \approx \rho_S$. At an intermediate distance, say at a distance of 10 m from the cylindrical axis, the object appears as an infinite cylinder with $|\mathbf{D}| \approx \rho_L/2\pi\rho = \rho_S 2\pi(0.5)/2\pi\rho = \rho_S/2\rho$. Finally, for far distances say 1000 m, the object appears so small that is behaves as a point charge of $2\pi(0.5)(100)\rho_S$ with $|\mathbf{D}| \approx 2\pi(0.5)(100)\rho_S/4\pi r^2 = 25/r^2$. $\mathbf{D}$ points away from the origin for all three field points.

These applications of Gauss' law are valid only if the dielectric material is homogeneous. This restriction can be relaxed somewhat to allow permittivities that vary in the same manner as allowed for variations in the charge distributions, i.e., radial for spherical, radial for cylindrical, and axially for Cartesian coordinate systems, respectively. However, throughout this text, we will concentrate upon materials that are uniform.

It is quite simple to calculate $\mathbf{E}$ from $\mathbf{D}$ by the constitutive relationship $\mathbf{D} = \varepsilon\mathbf{E}$. In addition, the voltage (or potential) at a point can be determined by defining the location of a zero potential and then calculating $\int \mathbf{E} \cdot \mathbf{dl}$. For spherical geometries, the zero or reference potential is defined at $r = \infty$, i.e., $V(\infty) = 0$. This enables the potential for a spherically symmetric charge distribution to be written as

$$V(r) - V(\infty) = V(r) = -\int_L \mathbf{E} \cdot \mathbf{dl}$$

$$= -\int_{r'=\infty}^{r} \frac{Q_{ENC}}{4\pi\varepsilon r'^2}\mathbf{a}_r \cdot \mathbf{a}_r dr' = \frac{Q_{ENC}}{4\pi\varepsilon r}. \qquad (2.77)$$

The voltage of a spherically symmetric charge distribution varies as $1/r$. For cylindrical geometries, the zero potential is chosen at $\rho = 1$, i.e., $V(1) = 0$. The potential is given by

$$V(\rho) - V(1) = V(\rho) = -\int_L \mathbf{E} \cdot \mathbf{dl}$$

$$= -\int_{\rho'=1}^{\rho} \frac{\rho_L}{2\pi\varepsilon\rho'}\mathbf{a}_\rho \cdot \mathbf{a}_\rho d\rho' = \frac{\rho_L}{2\pi\varepsilon}\ln\rho. \qquad (2.78)$$

The potential of infinite, cylindrical charge distributions varies according to the $\ln\rho$. Finally, the zero potential for planar geometries is chosen as the plane of symmetry. As in Example 2.13-3, where the plane of symmetry is $z = 0$, the potential is calculated as

$$V(z) - V(0) = V(z) = -\int_L \mathbf{E} \cdot \mathbf{dl}$$

$$= -\int_{z=0}^{z} \frac{\rho_S}{\varepsilon}\mathbf{a}_Z \cdot \mathbf{a}_Z dz = \frac{z\rho_S}{\varepsilon}, \qquad (2.79)$$

the potential of a planar charge distribution increases linearly with distance from the surface on which the charge is located. More sophisticated calculations of potential are deferred to a more advanced course.

Example 2.13-5. Calculate the capacitance/meter of twin-lead, TV transmission line with conductors of 2a diameter separated by distance 2d. Neglect the effects of the dielectric spacing material. The diagram of the model for such a transmission line is shown in Fig. 2.23.

This is a challenging problem to solve exactly, but we can obtain an approximate solution with the aid of Gauss' law. Assume that there are equal and opposite line charge densities on the two PEC conductors, $\pm\rho_L$. In fact, the charge distributions are not quite uniform since the attraction of the opposite charges tends to produce a slight concentration on the facing surfaces. However, for the electrode spacing $d \gg a$, this effect is minimal.

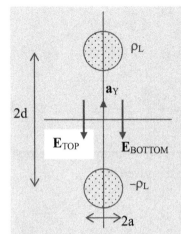

FIGURE 2.23: Twin-lead transmission line.

From Gauss law, the electric field along the y-axis due to the charge on the top conductor is

$$\mathbf{E}_{\text{TOP}} = \frac{\mathbf{D}_{\text{TOP}}}{\varepsilon} = \frac{\rho_L}{2\pi\varepsilon\rho_T}\mathbf{a}_{\rho\text{TOP}} = \frac{-\rho_L}{2\pi\varepsilon(d-y)}\mathbf{a}_Y$$

and the electric field due to the charge on the bottom conductor is

$$\mathbf{E}_{\text{BOT}} = \frac{\mathbf{D}_{\text{BOT}}}{\varepsilon} = \frac{\rho_L}{2\pi\varepsilon\rho_B}\mathbf{a}_{\rho\text{BOT}} = \frac{-\rho_L}{2\pi\varepsilon(d+y)}\mathbf{a}_Y.$$

Note that any path between the two conductors can be used for calculating the voltage, so we have chosen the convenient straight line of the y-axis. The voltage drop is found readily as

$$V_C = V_{\text{TOP}} - V_{\text{BOT}} = -\int (\mathbf{E}_{\text{TOP}} + \mathbf{E}_{\text{BOT}}) \cdot \mathbf{d}\ell$$

$$= -\frac{\rho_L}{2\pi\varepsilon}\int_{y-d+a}^{d-a}\left(\frac{1}{d+y}+\frac{1}{d-y}\right)(-\mathbf{a}_Y)\cdot\mathbf{a}_Y\,dy$$

$$= \frac{\rho_L}{2\pi\varepsilon}2\ln\left(\frac{2d-a}{a}\right) = \frac{\rho_L}{\pi\varepsilon}\ln\left(\frac{2d-a}{a}\right).$$

In general, the capacitance/meter of the twin-lead line is given as

$$C = \frac{\rho_L}{V_C} = \frac{\pi\varepsilon}{\ln\left(\frac{2d-a}{a}\right)} \approx \frac{\pi\varepsilon}{\ln\left(\frac{2d}{a}\right)}.$$

For twin-lead TV transmission line, the approximate dimensions are $2a = 1$ mm in diameter with a separation between wire axes of $2d = 1$ cm and $\varepsilon \approx 1.2\varepsilon_o$ which leads to

$$C \approx \frac{\pi\varepsilon}{\ln\left(\frac{2d}{a}\right)} = \frac{1.2(8.854)\pi}{\ln\left(\frac{0.01}{0.0005}\right)} = 11.1 \text{ pF/m}.$$

2.14   ARBITRARY CHARGE DISTRIBUTIONS

The unique configuration of the charge distributions of the previous examples allowed us to make many simplifications in the evaluation of Gauss' law. But, can we calculate the electric flux density for arbitrary distributions of charge? Most definitely, as long as the material surrounding the charges is linear and homogeneous. Most simply, the concept of linearity when applied to fields means that the field with two or more sources present is the vector sum of the fields

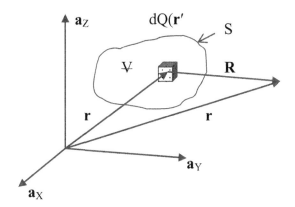

FIGURE 2.24: Electric flux density for arbitrary charge distribution.

of the individual sources acting alone. The total electric flux is found by adding the vector contributions of small, differential or incremental volumes of charge, each of which can be treated as a point charge. We will obtain the general form for the charge distribution shown in Fig. 2.24.

The volume charge distribution, located relative to an arbitrary coordinate origin O, varies according to position, i.e., $\rho_V = \rho_V(\mathbf{r}')$. $\mathbf{r}'$ is a shorthand way of indicating the coordinate position (x', y', z') where the prime denotes the position of an element of charge, frequently called the *source point*. The electric flux density is observed at the location $P(x, y, z)$ denoted by $\mathbf{r}$ and called the *field point*. This source-field point notation is commonly used in electromagnetics—$\mathbf{r}'$ and $\mathbf{r}$ as the source and field points, respectively.

A review of Eq. (2.76) shows that $\mathbf{D}$ associated with a point charge is directed radially outward and decreases as the square of the radial distance between the charge and the observation point. Since it occupies only a differential volume dv', the differential charge dQ acts as a point charge source. The radial vector from the source point to the observation, or field, point is expressed as

$$\mathbf{R} = \mathbf{r} - \mathbf{r}'. \tag{2.80}$$

The differential volume of charge at $\mathbf{r}'$ establishes a differential electric flux at $\mathbf{r}$ given by

$$d\mathbf{D}(\mathbf{r}) = \frac{dQ(\mathbf{r}')}{4\pi R^2}\mathbf{a}_R = \frac{\rho_V(\mathbf{r}')dv'}{4\pi R^2}\mathbf{a}_R \tag{2.81}$$

where $R = |\mathbf{R}| = |\mathbf{r} - \mathbf{r}'|$, dv' is the differential volume at $\mathbf{r}'$ and $\mathbf{a}_R = \mathbf{R}/|\mathbf{R}|$ is the unit vector in the direction from the source point to the field point. By integrating the contributions of all differential sources, we can find the electric flux due to the entire charge distribution as

$$\mathbf{D}(\mathbf{r}) = \frac{1}{4\pi} \iiint_V \left(\frac{\rho_V(\mathbf{r}')dv'}{R^2}\mathbf{a}_R \right). \tag{2.82}$$

For numeric evaluation, the incremental form becomes

$$\mathbf{D}(\mathbf{r}) = \frac{1}{4\pi} \sum_{i=1}^{N} \frac{\rho_V(\mathbf{r}_i')\Delta v_i'}{R_i^2} \mathbf{a}_{Ri} \qquad (2.83)$$

where the subscript i denotes the variable for the ith incremental volume. These forms express the vector superposition of "point-like" charges for a known distribution ρ_V. These equations can be modified for surface or line distributions. For surface distributions $dQ = \rho_s da$ and the flux density is given as

$$\mathbf{D}(\mathbf{r}) = \frac{1}{4\pi} \iint_S \frac{\rho_S(\mathbf{r}')\mathbf{a}_R}{R^2} da'. \qquad (2.84)$$

For numeric evaluation, the incremental form becomes

$$\mathbf{D}(\mathbf{r}) = \frac{1}{4\pi} \sum_{i=1}^{N} \frac{\rho_S(\mathbf{r}_i')\Delta a_i'}{R_i^2} \mathbf{a}_{Ri}. \qquad (2.85)$$

For line charge distributions, $dQ = \rho_L dL$, the flux density is given as

$$\mathbf{D}(\mathbf{r}) = \frac{1}{4\pi} \int_L \frac{\rho_L(\mathbf{r}')\mathbf{a}_R}{R^2} dL'. \qquad (2.86)$$

For numeric evaluation, the incremental form becomes

$$\mathbf{D}(\mathbf{r}) = \frac{1}{4\pi} \sum_{i=1}^{N} \frac{\rho_L(\mathbf{r}_i')\Delta L_i'}{R_i^2} \mathbf{a}_{Ri}. \qquad (2.87)$$

Of course, Eq. (2.83) is applicable for point charges where the dQ is replaced by the charge of each point charge

$$\mathbf{D}(\mathbf{r}) = \frac{1}{4\pi} \sum_{i=1}^{N} \frac{q_i}{R_i^2} \mathbf{a}_{Ri}. \qquad (2.88)$$

Equations (2.82) and (2.83) assume that the charge distribution is known. More often, the charge distribution is unknown and the voltage differences between various electrodes are known. Advanced methods beyond the scope of this text are used in these cases.

The compact and simple form of Eqs. (2.82) and (2.83) is deceptive since their evaluation is often rather complicated. The problem is primarily due to $\mathbf{a}_R$, the unit vector which points from the source to the field point, has a different direction for each source point. As you might expect, this makes things get pretty nasty. Rather than learn a bunch of mathematical tricks for this evaluation, let's look at a single concept which always works—simply decompose $\mathbf{a}_R$ into its Cartesian components which point in the same direction for all source locations. However, there is a price to be paid for this simplification. There are now three integrals, one for each Cartesian unit vector. The resulting three scalar integrals are additionally complicated due to the functional dependence of $\mathbf{a}_R$ on each of the Cartesian unit vectors.

Finally, calculation of $\mathbf{E}$ follows directly from the constitutive relationship $\mathbf{D} = \varepsilon\mathbf{E}$. As before, this requires that the permittivity is homogeneous.

Example 2.14-1. Calculate the flux density due to a uniform, infinitely long filament of charge directly by integration. As in Example 2.13-2, the filament is aligned with the z-axis, see Fig. 2.25. Since the flux density shows no z-dependence, we find it most convenient to evaluate the flux density at $z = 0$. The field point is located in the $z = 0$ plane at a radius of ρ gives $\mathbf{r} = \rho\mathbf{a}_\rho = (x^2 + y^2)^{1/2}(\cos\phi\mathbf{a}_X + \sin\phi\mathbf{a}_Y) = x\mathbf{a}_X + y\mathbf{a}_Y$. The flux density has no angular dependence. The charge is located along the z-axis, $\mathbf{r}' = z'\mathbf{a}_Z$. The distance between the source and field points is $R = |\mathbf{R}| = |\mathbf{r} - \mathbf{r}'| = [(\mathbf{r} - \mathbf{r}') \cdot (\mathbf{r} - \mathbf{r}')]^{1/2} = [x^2 + y^2 + z'^2 + 2\mathbf{r} \cdot \mathbf{r}']^{1/2} = [x^2 + y^2 + z'^2 + 2\rho\mathbf{a}_\rho \cdot z'\mathbf{a}_Z]^{1/2} = [x^2 + y^2 + z'^2]^{1/2}$. The unit vector $\mathbf{a}_R = \mathbf{R}/R = [x\mathbf{a}_X + y\mathbf{a}_Y - z'\mathbf{a}_Z]/[x^2 + y^2 + z'^2]^{1/2}$. Combining these factors in Eq. (2.86), we obtain

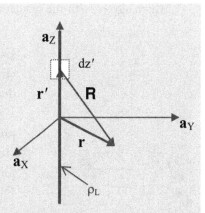

FIGURE 2.25: Electric flux density calculations for an infinite filament.

$$\mathbf{D}(\mathbf{r}) = \frac{1}{4\pi} \int_{z'=-\infty}^{\infty} \frac{\rho_L(x\mathbf{a}_X + y\mathbf{a}_Y - z'\mathbf{a}_Z)\,dz'}{[x^2 + y^2 + z'^2]^{3/2}}.$$

The integral of the z-component is an odd function of z' and as such vanishes over equal and opposite limits. This means that there is no z-component of the flux; all the flux is directed outward from the filament of charge. The integrals associated with the x and y components are even functions of z' and with similar forms. From tables or via MAPLE™, the integral is evaluated as

$$\int = \int_{z'=-\infty}^{\infty} \frac{dz'}{[a^2 + z'^2]^{3/2}} = 2\int_{z'=0}^{\infty} \frac{dz'}{[a^2 + z'^2]^{3/2}}$$

$$= \left.\frac{2z'}{a^2\left(a^2 + z'^2\right)^{1/2}}\right|_{z'=0}^{\infty} = \frac{2}{a^2}.$$

The electric flux density is given by

$$\mathbf{D(r)} = \frac{2\rho_L}{4\pi} \frac{(x\mathbf{a}_X + y\mathbf{a}_Y)}{[x^2 + y^2]} = \frac{\rho_L}{2\pi} \frac{\mathbf{a}_\rho}{[x^2 + y^2]^{1/2}}$$

$$= \frac{\rho_L}{2\pi\rho}\mathbf{a}_\rho,$$

which agrees with the results found directly by Gauss' law.

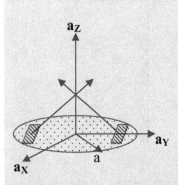

FIGURE 2.26: Electric flux density calculation for a disk.

Example 2.14-2. Calculate the field along the axis of a circular disk of radius $\rho = a$ with uniform surface charge density of ρ_S. For convenience, let's locate the disk in the $z = 0$ plane centered at the origin, see Fig. 2.26. Let's approach this problem with direct calculation of $\mathbf{D}$ from the charge distribution. The magnitude of the flux of a differential element of charge is given by $d|\mathbf{D}| = dQ/4\pi R^2$. From Fig. 2.26, it is obvious that the x and y components of the flux cancel leaving only the z-component which is expressed in terms of the geometry as $dD_z = d|\mathbf{D}|(z/R) = dQz/4\pi R^3 = z\rho_S\rho'd\rho'd\phi'/4\pi R^3$. In the same manner as in Example 2.14-1, $\mathbf{r} = z\mathbf{a}_Z$, $\mathbf{r}' = \rho'\mathbf{a}'_\rho$, and $R = [x'^2 + y'^2 + z^2]^{1/2}$. These details lead to

$$\mathbf{D(r)} = \frac{\mathbf{a}_Z}{4\pi} \int_{\rho'=0}^{a} \int_{\phi'=0}^{2\pi} \frac{\rho_S z\rho'd\rho'd\phi'}{[\rho'^2 + z^2]^{3/2}} = \frac{\mathbf{a}_Z 2\pi\rho_S z}{4\pi} \int_{\rho'=0}^{a} \frac{\rho'd\rho'}{[\rho'^2 + z^2]^{3/2}}$$

$$= \frac{\mathbf{a}_Z\rho_S}{2}\left(1 - \frac{z}{[a^2 + z^2]^{1/2}}\right).$$

Note that as $a \to \infty$, i.e., the disk becomes a plane, $D_z \to \rho_S/2$ as we found earlier for a uniform planar charge distribution. Also, as z gets very large, then $D_z \to (\rho_S/2)(a^2/2z^2) = \pi a^2\rho_S/4\pi z^2 = Q_S/4\pi z^2$ for finite a, the flux density of a point charge.

There is so much more to say about the fields due to charges. But, there is so little time and so many more subjects to study. You can extend your studies by enrolling in advanced electromagnetics courses.

CHAPTER 3

Inductors

3.1 INDUCTORS: A FIRST GLANCE

As you know from circuits, inductors are used primarily as a storage element for magnetic energy. Proper choice of materials and their configuration enhances this property and minimizes losses. Circuit theory has shown us that the terminal characteristics of inductors is given by

$$V_L = L\frac{dI_L}{dt} \qquad (3.1)$$

where I_L is the current flowing into the inductor in the direction of the voltage drop across the inductor, V_L, and L is the inductance value of the inductor expressed in Henries and abbreviated as H. Capacitors exhibited many similarities of structure and fields with those of resistors that allowed us to use many concepts developed for resistors in our understanding of capacitors. However, the dramatically different nature of magnetic fields requires several additional concepts to explain the behavior of inductors.

The properties of inductors depend upon a magnetic flux that exists mainly within the flux-guiding material much as in resistors and capacitors. As we will learn in the following sections, magnetic flux is the result of current flow. Consequently, metallic, current-carrying wires are usually used to establish magnetic flux. With many turns of wire wound into a helical coil, the fields of individual turns add together enabling a small current to produce a large magnetic flux. In contrast to current flux and electric flux that begin and end upon electrodes, magnetic flux forms a closed flux lines with no beginning or end. If the flux-guiding material does not provide a closed path, then the flux uses the air surrounding the coil as the guiding medium. This is inefficient and makes prediction of the field configuration difficult. These shortcomings are overcome by using a flux guide made of magnetic material. A common configuration for inductors is a closed flux guide of magnetic material (often called a *core*) with a coil of many turns wound upon it, see Fig. 3.1. The coil–core combination is often called a *magnetic circuit*. Though inductors come in many other shapes, we will focus on the coil–core inductor configuration through out this text.

The metallic wire leads of resistors and capacitors are relatively short and exhibit such small resistance that they are modeled as PEC wires. On the other hand, the wire of the coil can

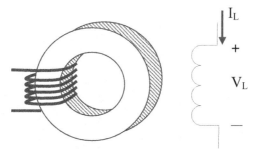

FIGURE 3.1: An inductor: (a) physical configuration and (b) electrical model.

be quite long and is likely to have much more resistance. Since we will limit ourselves to linear core material, the total voltage drop across the inductor is the sum of the resistive voltage drop and the voltage drop due to changing inductor current, see Eq. (3.1). An accurate model must include both of these effects. The resistance value can be easily calculated via Eq. (1.2) where L is the length of the wire and A is its cross-sectional area. However, for most of our work, we will focus upon the inductive effects; inclusion of the resistive term unnecessarily complicates our understanding the magnetics associated with inductors. Consequently, we will ignore the resistance of the wire, assuming it to be a PEC, see Fig. 3.1. Actually, this approximation is fairly accurate for many applications. If a more accurate model is needed, the resistance of the wire can be included as a resistor in series with the inductor.

Several simplifying assumptions regarding the magnetic flux guide are necessary in order that the mathematics remains reasonable. An accurate model for the magnetic flux guide is complicated by the fact that most magnetic materials are lossy and nonlinear. For simplicity, and without any significant alteration of fundamental behavior, we will consider lossless magnetic material only. Initially, we will assume linearity of the magnetic material as well. However, some of our work in Chapter 4 will involve nonlinear material. Finally, we will assume that the core is an ideal flux guide, i.e., no flux "leaks" out. As we will soon discover, this condition is met by almost all magnetic materials.

3.2 MAGNETIC FLUX DENSITY

Similar to current flux density and electric flux density, a *magnetic flux density*, **B**, is defined with units of webers/square meter [wb/m^2] in honor of Wilhelm Weber. Unfortunately, the SI system "masks" the fact that this is a flux density by the use of alternate units of Tesla T. The magnetic flux through a surface S is defined as

$$\Psi_m = \iint\limits_S \mathbf{B} \cdot \mathbf{ds} \ [\text{wb}] \qquad (3.2)$$

in webers. Though many textbooks use the symbol Φ to represent *magnetic flux*, we will continue the notation Ψ (with a subscript "m") that we began in the first two chapters. Sometimes you will see magnetic flux with units of lines, particularly in power related references.

So far our description of **B** seems much like **J** and **D**. On the other hand, experimental observations have never revealed magnetic charge. Consequently, there is no source from which magnetic lines emanate. Mathematically this is expressed as

$$\oiint_S \mathbf{B} \cdot \mathbf{ds} = 0 \tag{3.3}$$

This means that there is no net flux emanating from any arbitrary surface S including surfaces that become vanishingly small. No matter which region is enclosed, there are no magnetic charges inside. Since lines of flux begin or end on sources or sinks, Eq. (3.3) implies that magnetic flux lines have no beginning or end. Instead, they close upon themselves, see Fig. 3.2. This characteristic explains why magnetic fields are established without a pair of electrodes as required in resistors and capacitors. The current within the coil establishes the magnetic flux that passes through the coil and closes upon itself.

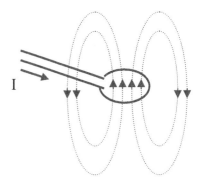

FIGURE 3.2: Magnetic flux through a coil.

Since Eq. (3.3) holds for all closed surfaces, we select a vanishingly small surface and apply the divergence theorem to obtain

$$\lim_{S \to 0} \oiint_S \mathbf{B} \cdot \mathbf{ds} = \iiint_\forall \nabla \cdot \mathbf{B} dv$$
$$\approx \nabla \cdot \mathbf{B} \Delta \forall = 0. \tag{3.4}$$

Since the volume $\Delta \forall \neq 0$, then

$$\nabla \cdot \mathbf{B} = 0, \tag{3.5}$$

the differential description of the continuity of magnetic flux density and is valid everywhere; there is no isolated magnetic charge anywhere. This type of field is called *divergenceless* since the divergence is identically zero.

This magnetic form of Gauss' law describes the closed nature of magnetic flux lines and the reason that a closed magnetic flux guide is desired. However, how is the magnetic field related to the current? This is the topic of the next section.

3.3   AMPERE'S LAW AND MAGNETIC FIELD INTENSITY

As with electric fields, we postulate that magnetic flux density is accompanied by a *magnetic field intensity*, **H**, with which magnetic phenomena can be correctly calculated. Since there is no evidence to refute this model for magnetic behavior, let's use it. Magnetic field intensity, **H**, with units of A/m, is established by current flow. Based on experimental evidence, André Ampere formulated the relationship between an electric current and the magnetic field intensity that it produces; this is known as Ampere's law and is expressed as

$$\oint_L \mathbf{H} \cdot \mathbf{dl} = I_{\text{ENC}}. \qquad (3.6)$$

The work integral of magnetic field intensity around a closed path is equal to the current enclosed by that path, see Fig. 3.3. The positive direction for the current, I_{ENC}, is defined as the upward direction as you "walk" around the path L in a direction so that the enclosed region is always on your left. Alternatively, this is expressed by the *right–hand rule*—when the fingers of your right hand point in the direction of integration along the path L, then your thumb points in the direction of the positive current flow. The closed paths L are called Amperian paths.

The enclosed current on the RHS of Eq. (3.6) can be written in several different forms representing filamentary, surface, and volume current flow. When the path L encloses several *filamentary currents*, see Fig. 3.3(a), the form of Eq. (3.6) becomes

$$\oint_L \mathbf{H} \cdot \mathbf{dl} = I_{\text{ENC}} = \sum_{\text{Wires}} I_{\text{Wires}} \qquad (3.7)$$

where the current is positive if it flows in the direction of the positive surface normal and negative if oppositely directed. Current-carrying wires are often modeled by filamentary currents. For *surface current densities*, see Fig. 3.3(b), Ampere's law becomes

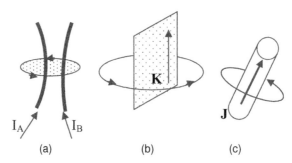

(a)   (b)   (c)

FIGURE 3.3: Geometry for Ampere's law: (a) filamentary currents, (b) surface currents, and (c) volume currents.

$$\oint_L \mathbf{H} \cdot \mathbf{dl} = I_{\text{ENC}} = \sum_i \int_{L_i} \mathbf{K}_i \cdot \mathbf{a}_N dl_i \qquad (3.8)$$

where integrals over all of the surface currents are added together. Of course, *volume current density* is also included as

$$\oint_L \mathbf{H} \cdot \mathbf{dl} = I_{\text{ENC}} = \iint_S \mathbf{J} \cdot \mathbf{ds} \qquad (3.9)$$

where the surface S has its perimeter defined by the path L and its normal is chosen by the right-hand rule stated earlier. Though there are an infinite number of possible surfaces S as shown in Fig. 3.3, it doesn't matter which one is chosen. Any surface for which the path L forms its edge will give the same result. Planar surfaces, as in Fig. 3.3(a), are used often, but the "bowl" form of the surface of Fig. 3.3(c) gives the same result. Only one direction of the surface normal from S is positive; current densities in the other direction contribute negatively to the RHS of Eq. (3.6). Line and surface integral evaluation techniques we have used earlier can be applied to the integrals on both sides of Eq. (3.6).

But, what is the meaning of Eq. (3.6)? The work integral of the magnetic field intensity around the closed path L is not zero as it is for the work integral of electric field intensity. Completion of a closed circuit around the path L requires no work for electric fields, but work is required for magnetic fields. This means that the magnetic field intensity is not a conservative field, another significant difference between the magnetic fields of inductors and the electric fields of resistors and capacitors.

A closer look at Eq. (3.6) and Fig. 3.3 reveals that the work integral around the path L is equal to the current flux penetrating the surface S. As long as there is no net current flux through the surface S, the work around the closed path L will be zero. The magnetic field intensity depends upon the current flux via this relationship. This connection between field intensity and flux density, as in Eq. (3.9), plays an important role in other aspects of electromagnetics as well.

Example 3.3-1. Calculate the closed integral of magnetic flux intensity, $\oint \mathbf{H} \cdot \mathbf{dl}$, around the several paths shown in Fig. 3.4. Path 1 is oriented in a CCW fashion so that positive current is directed out of the page. The dark circles represent current out of the plane of the paper; the cross represents current into the paper. The path encloses $+I_A$, $+I_B$, and $-I_C$ for a total current of $I_{\text{TOTAL}} = I_A + I_B - I_C = \oint \mathbf{H} \cdot \mathbf{dl}$. Path 2 is oriented in a CW fashion so that positive current is directed into the page. It encloses currents I_A and $-I_C$, but they contribute oppositely due to the right-hand rule. Therefore, $\oint \mathbf{H} \cdot \mathbf{dl} = -I_A + I_C$.

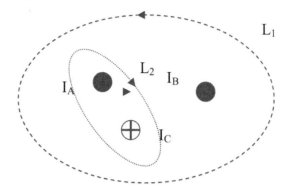

FIGURE 3.4: Amperian paths.

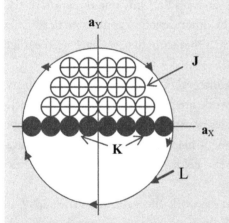

FIGURE 3.5: Surface and volume currents enclosed by circular path.

Example 3.3-2. A surface current density expressed as $\mathbf{K} = 10|x|\mathbf{a}_Z$ A/m flows on the $y = 0$ plane; a volume current density $\mathbf{J} = -10\mathbf{a}_Z$ A/m^2 exists in the positive y half-space, i.e., $y > 0$; see Fig. 3.5. Calculate the current enclosed by the CW circular path of radius 4 in the $z = 0$ plane.

By the right-hand (RH) rule, the positive direction for the closed path is in the negative z-direction, $-\mathbf{a}_Z$, or into the plane of the paper. The circular path with $r = 4$ encloses all of the surface current in the range $-4 \leq x \leq 4$ calculated via Eq. (3.8) as

$$I_1 = \int\limits_L \mathbf{K} \cdot \mathbf{a}_N dl = \int\limits_{x=-4}^{4} 10|x|\mathbf{a}_Z \cdot (-\mathbf{a}_Z)dx = -2(10)\frac{16}{2} = -160 \text{ A}.$$

The path encloses volume distributed current for $y > 0$ calculated via Eq. (3.9) as

$$I_2 = \iint\limits_S \mathbf{J} \cdot \mathbf{ds} = \int\limits_{\phi=0}^{\pi} \int\limits_{\rho=0}^{4} (-10\mathbf{a}_Z) \cdot (-\mathbf{a}_Z)\rho d\rho d\phi = 10\pi\frac{16}{2} = 251.3 \text{ A}.$$

This leads to $I_{\text{TOTAL}} = I_1 + I_2 = -91.3$ A.

3.4 MAGNETIC FIELDS IN CORES

Ampere's law provides considerable insight into the global behavior of magnetic fields. In general it has conceptual value, but is not very useful for general calculations. However, when applied to a *closed magnetic flux guide*, it is a useful calculation tool. Closed magnetic flux guide is a rather long and imposing name so let's use the more common name of *magnetic core* or just *core*. To make our work easier, we will assume that no magnetic flux leaks from the core, i.e., it is an ideal flux guide.

Experimental evidence has shown that when a coil is wound on a core, the magnetic field is well defined by Ampere's law and a form of the right-hand rule. Consider the toroidal core shown in Fig. 3.6(a) with a uniform cross section and excited by an N-turn coil of a wire carrying a current of I amperes. The size of the wire has only a secondary effect upon flux leakage from between the turns (also known as fringing); for simplicity, we assume that the wire is filamentary and there is no leakage flux. However, keep in mind that when fabricating such an inductor, the wire size does limit the number of turns that can fit through the center of the core. Finally, the location of the coil on a branch of the core has very little effect on the field induced in that branch.

An alternative statement of the *right-hand rule* states that when the fingers of the right hand are oriented in the direction of current flow, the thumb points in the direction of magnetic field induced in the core, see Fig. 3.6(b). This means that the field in Fig. 3.6(a) is directed angularly through the core in a counter-clockwise (CCW) direction and the magnetic field is expressed as $\mathbf{H} = H_\phi \mathbf{a}_\phi$. Experimental evidence has shown that due to the symmetry of toroidal cores with rectangular cross section, the field depends only upon radial position within the core, i.e., $H_\phi(\rho, \phi, z) = H_\phi(\rho)$. Before applying Ampere's law, select a path aligned with the magnetic field, see Fig. 3.6(c). A circular path of radius ρ is chosen for L. When these

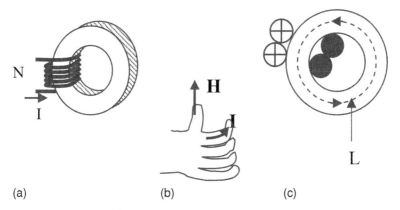

(a) (b) (c)

FIGURE 3.6: Magnetic circuit: (a) circuit diagram, (b) RH rule, and (c) integration path.

conditions are applied to Ampere's law, the result is

$$\oint_L \mathbf{H} \cdot \mathbf{dl} = \int_{\phi=0}^{2\pi} H_\phi \mathbf{a}_\phi \cdot \mathbf{a}_\phi \rho d\phi$$

$$= \rho H_\phi \int_{\phi=0}^{2\pi} d\phi = 2\pi \rho H_\phi = I_{ENC}. \qquad (3.10)$$

According to the convention for relating line and surface integrals, when the fingers of the right hand are directed along the path of integration, the thumb points upward in the direction of positive current. The path L encloses the currents that flow upward in the part of each loop along the inner surface of the toroid. The current flowing on the other three sides of each loop are not enclosed by the path L. The enclosed current comprises N separate filaments each with current I so that $I_{ENC} = NI$ and the magnetic field intensity is given as

$$H_\phi = \frac{NI}{2\pi\rho}. \qquad (3.11)$$

Note that with the positive z-direction upward out of the plane of the paper the phi-direction is CCW. The magnetic field at the outer radius of the toroid is smaller than the field at the inner radius due to the inverse radial dependence of the path length, $1/\rho$.

We have obtained an expression for the magnetic field intensity. But, before we can proceed further in describing the behavior of the magnetic fields in the core, we must relate the field intensity to the flux density. As we saw with resistors and capacitors, the relationship between field intensity and flux density is fairly simple and depends on material properties. The next section considers this relationship.

Example 3.4-1.   Calculate the maximum and minimum values of the magnetic field intensity within the core of Fig. 3.6(a) for a toroidal core with an inner radius of 3 cm, an outer radius of 4 cm, a thickness of 1 cm, and with a 50-turn coil that carries a current of 5 A. From Eq. (3.11), the field is given as $H_\phi = \frac{NI}{2\pi\rho} = \frac{50(5)}{2\pi\rho} = \frac{39.8}{\rho}$ A/m. The requested values are $H_{\phi MIN} = 994.7$ A/m at the outer edge of the toroid and $H_{\phi MAX} = 1326.3$ A/m at the inner edge.

Example 3.4-2.   Calculate the current required for a minimum field intensity of 1000 A/m within a toroid core with an inner radius of 2 cm, an outer radius of 5 cm, and a thickness of 1 cm with a 100-turn coil. The minimum field occurs at the outer edge of the toroid at $\rho = 0.05$ so that the required current is given by $I = \frac{2\pi\rho H_\phi}{N} = \frac{2\pi(0.05)1000}{100} = 3.14$ A.

3.5   MAGNETIC PERMEABILITY

Magnetic material is characterized by a constitutive property known as *magnetic permeability* or just *permeability*. This parameter "connects" the field intensity and flux density of magnetic material much as conductivity and permittivity connected the field intensity and flux density of resistors and capacitors, respectively. Permeability is denoted by the symbol μ and has units of Henries/meter (H/m) in the SI system. The usual form for defining permeability is

$$\mathbf{B} = \mu\mathbf{H}. \tag{3.12}$$

Actually, the relationship is much more complicated than a scalar constant. Typically it includes losses, nonlinearity, and hysteresis. In some cases it has a tensor nature. We can ignore these effects most of the time without significant loss of accuracy. However, we will consider briefly the simpler aspects of nonlinearity in a later chapter.

Magnetic properties are based upon atomic behavior. The simplest model for this behavior consists of a neutral atom with a central positive nucleus surrounded by a circulating negative electron cloud. The orbital motion of the electrons produces a circular current that encircles the central atom. By the right-hand rule, the magnetic field established by this current passes upward through the center of the atom as shown in Fig. 3.7(a). For simplicity, a single circulating charge represents the entire electron cloud. In the absence of an external field, thermal energy keeps the atom and its magnetic field randomly oriented. But, when an external magnetic field is applied, it interacts with the circulating current of the atom, tending to align the field of the atom with the applied field by "tipping" the plane of the circulating cloud charge, see Fig. 3.7(b). This alignment extracts energy from the external magnetic field, storing it as mechanical potential energy. In reality, a group of atoms with aligned fields form a *domain*; in the absence of an external field, the domains are randomly oriented and their fields cancel each other. With the application of an external magnetic field, domains in the direction of the field grow in size, while all others shrink. The result is a significant increase in the flux within the material due to the alignment with the external field and the fields of nearly all the atoms in the material. In a sense the magnetic material acts as a flux multiplier, increasing it significantly over that of free space. With the removal of the external field, the domains become nearly uniform in size and random in orientation due to thermally induced vibrations; their magnetic fields essentially cancel each other.

This model helps to explain the absence of isolated magnetic charges or poles. The fields generated by the circulating electron clouds cannot be divided into outward directed flux and inward directed flux as is possible with electric flux due to charges. Instead, the inward and outward directed fluxes are inseparable; the outward flux is a continuation of the inward flux. The continuous flux lines penetrate the circulating electron cloud.

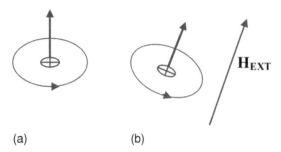

(a)   (b)

FIGURE 3.7: Atomic circulating currents and magnetic field: (a) without external field and (b) with external field.

At low levels of **H**, increases of I_L and the corresponding **H** are accompanied by great increases in **B** as the domains in the direction of the field grow rapidly. At higher levels, most of the domains are already aligned and much smaller changes in **B** occur for the same increases in I_L and **H**. **B** becomes saturated at high levels of I_L and **H**. This nonlinear relationship between **B** and **H** is frequently displayed in so-called magnetization or B–H curves, see Fig. 3.8. The exact shape and magnitude of the B–H curve depends upon the magnetic material.

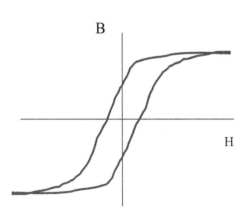

FIGURE 3.8: Magnetization or B–H curve.

As **H** increases, **B** increases as favorable domains expand. But the motion is not smooth and continuous. Rather, the domain walls tend to "stick" together with no motion as **H** increases slightly; then a very slight additional increase in **H** causes a sudden jump in the domain wall. This process is accompanied by energy loss. For reductions of **H**, the effect is similar, but **B** doesn't decrease as rapidly as **H** does. This effect is known as *hysteresis*. A typical B–H curve with hysteresis for successively larger cycles of excitation is shown in Fig. 3.8.

Permanent magnets are made from alloys of iron, nickel, cobalt, aluminum, and samarium. While in the molten state, the material is placed in a strong external magnetic field that produces large domains aligned with the external field. Then the magnet is slowly cooled, "freezing in" the aligned domains. When cooled to room temperature, the external magnetic field is removed, yet the field due to the permanently aligned domains remains.

Magnetic materials become increasingly lossy as the signal frequency is increased above a few tens of Hertz. However, specially fabricated, low-loss, ferrite materials operate satisfactorily up to 1 GHz.

The permeability of free space is denoted by μ_o and is equal to $4\pi \times 10^{-7}$ H/m in the SI system. Materials are often described in terms of relative permeability which is defined as

$$\mu_R = \frac{\mu}{\mu_o}. \qquad (3.13)$$

Nonmagnetic materials are described by $\mu_R = 1$. Magnetic materials commonly have values of μ_R on the order of 10^3, some even as large as 10^5. As we will see in later sections, materials with such large values of μ_R approach the performance of an ideal flux guide with little or no leakage flux.

Example 3.5-1. Calculate the total flux in the core of Example 3.4-1 with $\mu_R = 1000$. The magnetic flux density is calculated with Eq. (3.12) as

$$B_\phi = \mu H_\phi = \frac{\mu NI}{2\pi\rho} = \frac{10^3(4\pi \times 10^{-7})39.8}{\rho} = \frac{50}{\rho} \text{ mwb/m}^2.$$

The total flux is calculated by Eq. (3.2) as

$$\Psi_m = \iint_S \mathbf{B} \cdot \mathbf{ds} = \int_{z=0}^{0.01} \int_{\rho=0.03}^{0.04} \frac{0.05}{\rho} d\rho dz = 0.05(0.01)\ln\left(\frac{0.04}{0.03}\right) = 0.14 \text{ mwb}.$$

3.6 MAGNETIC BOUNDARY CONDITIONS

By now you know the procedures for determining boundary conditions. A closed surface integral is evaluated to determine the flux density boundary condition. A closed line integral provides the boundary condition for the field intensity. Let's consider the flux density first.

Following the development used in Sections 1.25 and 2.6, we substitute $\mathbf{B}$, magnetic flux density, for $\mathbf{D}$, electric flux density, in Eq. (2.16). In addition, we include the fact that there is no magnetic charge, see Fig. 3.9. The resulting boundary condition is

$$B_{N1} = B_{N2} \qquad (3.14)$$

or in vector form

$$\mathbf{a}_N \cdot (\mathbf{B}_2 - \mathbf{B}_1) = 0. \qquad (3.15)$$

The flux density is continuous throughout magnetic material and at all boundaries. This is ensured by the absence of magnetic charge. This behavior is often noted as the *continuity of magnetic flux*.

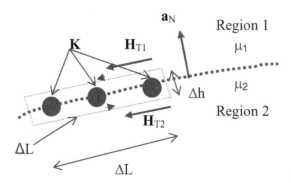

FIGURE 3.9: Magnetic field intensity boundary conditions.

Using the same procedure as with electric fields, we find that the boundary condition for **H** is slightly different than for **E**. A long, narrow closed path L is centered on the boundary, see Fig. 3.9. Evaluating Ampere's law on this boundary using a CCW path, we find that the LHS becomes

$$\oint_L \mathbf{H} \cdot \mathbf{dl} \approx (H_{N1R} + H_{N2R} - H_{N1L} - H_{N2L})\left(\frac{\Delta h}{2}\right) + (H_{T1} - H_{T2})\Delta L. \qquad (3.16)$$

In the limit, as $\Delta h \ll \Delta L \to 0$, the Δh term becomes insignificant and we have

$$\oint_L \mathbf{H} \cdot \mathbf{dl} \approx (H_{T1} - H_{T2})\Delta L = I_{ENC}. \qquad (3.17)$$

Due to the nonconservative nature of magnetic fields, the RHS is not automatically zero as it was for electric fields. As the width of the path shrinks to zero, no current will be enclosed by the path L except surface current, **K**, since the loop is always centered on the boundary. The total current enclosed is given by

$$I_{ENC} = \int_{\Delta L} \mathbf{K} \cdot \mathbf{a}_{OUT} dl = K_{OUT}\Delta L \qquad (3.18)$$

where K_{OUT} is the component of **K** that points perpendicularly out of the page and is enclosed by L. Equating Eqs. (3.17) and (3.18), we obtain the boundary condition on **H** as

$$H_{T1} - H_{T2} = K_{OUT} \qquad (3.19)$$

or in vector format as

$$\mathbf{a}_N \times (\mathbf{H}_1 - \mathbf{H}_2) = \mathbf{K} \qquad (3.20)$$

where $\mathbf{a}_N$ points from region 2 to region 1. The magnetic field intensity is discontinuous at a boundary by an amount equal to the surface current density on that boundary. The nonconservative nature of the magnetic field explains the nonzero term on the RHS of Eqs. (3.19) and (3.20). The vector nature of this boundary condition is due to the complex relationship between electric currents and the magnetic fields that they establish. We will learn more of this later.

The development of magnetic boundary conditions to this point is completely general, applicable at all boundaries. However, in many DC and low-frequency applications, surface currents do not occur naturally at magnetic–magnetic boundaries. In these cases, the RHS of Eqs. (3.19) and (3.20) are zero and boundary conditions take on the same form as those of electric fields, i.e., $H_{T1} = H_{T2}$. For this case, we can combine Eqs. (3.13) and (3.19) with $\mathbf{B} = \mu \mathbf{H}$ to express the magnetic–magnetic boundary conditions exclusively in terms of $\mathbf{B}$ or $\mathbf{H}$ as

$$B_{N1} = B_{N2} \quad \text{and} \quad \frac{B_{T1}}{\mu_1} = \frac{B_{T2}}{\mu_2}$$

$$\mu_1 H_{N1} = \mu_2 H_{N2} \quad \text{and} \quad H_{T1} = H_{T2}. \tag{3.21}$$

These compare with Eqs. (1.100) and (2.23) for conductive and dielectric materials, respectively.

A few words about magnetic fields in metal conductors are in order. DC magnetic fields in metals are rarely accompanied by naturally occurring surface currents and so they satisfy the boundary conditions of Eq. (3.21). *Since metals have $\mu_r = 1$, DC magnetic fields are unaffected by their presence.* On the other hand, AC magnetic fields *within* metals tend toward zero while the currents in the metals tend to be located near the surface of metals. The higher the frequency and the greater the conductivity, the more pronounced are these tendencies. Consequently, for high frequency fields near good conductors, a good approximation is to assume that the fields are zero within the metal and that the current is entirely on the surface. This is equivalent to approximating the metal as a PEC. When the material in region 1 is a PEC, the magnetic boundary conditions become

$$\mathbf{a}_N \cdot \mathbf{B}_2 = 0 \quad \text{and} \quad \mathbf{a}_N \times \mathbf{H}_2 = -\mathbf{K}. \tag{3.22}$$

No AC magnetic flux penetrates into the PEC and the magnetic field adjacent to the PEC is tangential and equal to the surface current density on the PEC.

We have assumed that the core acts as an ideal flux guide, i.e., no flux leaks out of the core. Actually, there is some leakage, but the smaller it is, the more accurate our approximation. As with resistors and capacitors, flux leakage is related to the material properties at the surfaces of the core. From Eq. (3.21), we see that the tangential fluxes are related by $B_{T2} = (\mu_2/\mu_1)B_{T1}$. When we assign region 1 as the magnetic core with $\mu_1 \gg \mu_o$ and region 2 as air with $\mu_2 = \mu_o$,

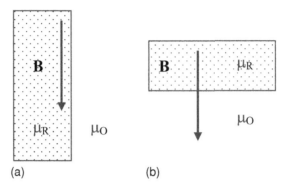

FIGURE 3.10: Magnetic flux at boundaries: (a) tangential flux and (b) perpendicular flux.

then $\mu_2/\mu_1 = 1/\mu_r \ll 1$ and there is very little tangential flux adjacent to the core. Under these conditions, the core behaves nearly as an ideal flux guide. For flux that is perpendicular to a boundary as in an air gap in the core, Eq. (3.21) shows that $B_{N1} = B_{N2}$. This means that all of the flux normal to a boundary extends from one material to the other.

For closed magnetic cores with the magnetic field established by current in a coil wound on the core, the flux is essentially tangential to the surfaces of the core. Since $\mu_r \gg 1$ for most magnetic materials, there is very little flux leakage; this situation is accurately approximated as an ideal flux guide, see Fig. 3.10(a). Many applications such as rotating motors require a gap in the magnetic material that is perpendicular to the flux. In these situations, all of the flux within the core crosses the gap since $B_{N1} = B_{N2}$, see Fig. 3.10(b).

Example 3.6-1.   A magnetic material with relative permeability $\mu_R = 2$ occupies the region for $x \le 0$ and is bounded by air at the $x = 0$ plane. The magnetic field within the magnetic material at the boundary is given by $\mathbf{H}_2 = 5\mathbf{a}_X + 5\mathbf{a}_Y + 5\mathbf{a}_Z$ A/m. No surface current flows on the boundary. Calculate $\mathbf{H}_1$ and $\mathbf{B}_1$ in the air adjacent to the boundary. If the normal to the boundary is defined as pointing from region 2 to region 1, $\mathbf{a}_N = \mathbf{a}_X$, then $H_{N2} = 5$ and $H_{T2} = 5\sqrt{2}$ A/m with a direction of $\mathbf{a}_T = 0.707(\mathbf{a}_Y + \mathbf{a}_Z)$. In addition, $B_{N2} = 2\mu_o 5 = 10\mu_o$ and $B_{T2} = 10\sqrt{2}\mu_o$ wb/m^2. From Eq. (3.21), we obtain $H_{T1} = H_{T2} = 5\sqrt{2}$ A/m with $B_{T1} = \mu_o H_{T1} = 5\sqrt{2}\mu_o$ wb/m^2 and $B_{N1} = B_{N2} = 10\mu_o$ wb/m^2 with $H_{N1} = B_{N1}/\mu_o = 10$ A/m. Thus, the requested fields are $\mathbf{H}_1 = 10\mathbf{a}_X + 5\sqrt{2}(0.707)(\mathbf{a}_Y + \mathbf{a}_Z) = 10\mathbf{a}_X + 5(\mathbf{a}_Y + \mathbf{a}_Z)$ A/m and $\mathbf{B}_1 = 10\mu_o\mathbf{a}_X + 5\sqrt{2}\mu_o(0.707)(\mathbf{a}_Y + \mathbf{a}_Z) = 10\mu_o\mathbf{a}_X + 5\mu_o(\mathbf{a}_Y + \mathbf{a}_Z) = \mu_o\mathbf{H}_1$ wb/m^2.

Example 3.6-2.   Consider Example 3.6-1 with the addition of a surface current density of $\mathbf{K} = 5\mathbf{a}_Y$ A/m on the magnetic material–air interface. The normal component is unaffected by **K** so that $B_{N1} = B_{N2} = 10\mu_o$ wb/m^2 and $H_{N2} = 10$ A/m. The tangential components satisfy Eq. (3.20) as

$$\mathbf{K} = 5\mathbf{a}_Y = \mathbf{a}_N \times (\mathbf{H}_1 - \mathbf{H}_2)$$
$$= \mathbf{a}_X \times [(H_{X1} - 5)\mathbf{a}_X + (H_{Y1} - 5)\mathbf{a}_{Y2} + (H_{Z1} - 5)\mathbf{a}_Z]$$
$$= (H_{Y1} - 5)\mathbf{a}_Z - (H_{Z1} - 5)\mathbf{a}_Y$$

from which $H_{Y1} = 5$ and $H_{Z1} = 10$ A/m or $H_{T1} = 5\mathbf{a}_Y + 10\mathbf{a}_Z$ so that $B_{T1} = 5\mu_o\mathbf{a}_Y + 10\mu_o\mathbf{a}_Z$. The requested fields are $\mathbf{H}_1 = 10\mathbf{a}_X + 5\mathbf{a}_Y + 10\mathbf{a}_Z$ A/m and $\mathbf{B}_1 = 10\mu_o\mathbf{a}_X + 5\mu_o\mathbf{a}_Y + 10\mu_o\mathbf{a}_Z$ wb/m^2.

3.7 FARADAY'S LAW

The concepts of Ampere's law, continuity of magnetic flux, and permeability developed in the previous sections are applicable for DC currents. They can also be applied to time-varying fields with "low frequencies" with slight modifications. But an additional feature of time-varying magnetic fields known as Faraday's law is even more important—the generation of an electric field and an accompanying voltage by a changing magnetic field. Hans Christian Oersted's discovery[*] of a magnetic field generated by a current prompted Michael Faraday to seek the establishment of a current by a magnetic field. He was unsuccessful in demonstrating that a current is generated by a *steady* magnetic field. However, he discovered that a *changing* magnetic field generates an electric field and, hence, a voltage. He observed that a changing magnetic field induced a voltage in a nearby loop of wire according to

$$V_{\text{LOOP}} = -\frac{d\Psi_m}{dt} \tag{3.23}$$

where V_{LOOP} is the voltage measured across the ends of a wire loop and Ψ_m is the magnetic flux enclosed by the loop. When the voltage is due to changing magnetic flux it goes by several names: *electromotive force, emf*, or *induced voltage*. The induced voltage is equal to the negative rate of change of the magnetic flux through the loop. A nonvarying magnetic field will not induce an emf, only a changing field can do it. Initially, the minus sign seems unnecessary since we will use an alternate means to determine the polarity of the induced voltage. However, its inclusion ensures a consistency between Eq. (3.23) and a vector field representation of Faraday's law that we will cover later. Let's carefully consider several important aspects of this definition of emf.

---

[*] Professor Oersted observed the deflection of a magnetic compass needle by a nearby steady current during a classroom demonstration.

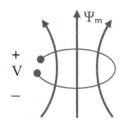

FIGURE 3.11: Geometry for simple application of Faraday's law.

The two ends of the wire loop are assumed to be infinitesimally close together and the leads of the meter used to measure the voltage are of negligible length. This means that a closed loop of wire is cut at a point and a voltmeter of vanishingly small size is connected to the two ends of wire. The voltmeter is assumed to be ideal, i.e., it has infinite impedance, so that no current flows through it. The wire loop is assumed to be a PEC so that resistive effects can be neglected. The polarity of the induced voltage requires a bit of explanation as well.

The basic principle that assists us in determining the polarity of the induced voltage is inertia. In electromagnetics, as in mechanics, nature resists changes from its present state. Henri Frederic Emile Lenz reasoned that *the induced voltage tends to produce a current that will flow in the direction that produces a flux that opposes the original change of flux*, hence, reducing the original induced voltage. This action tends to bring the electromagnetic state of the loop back to its original condition. This is known as *Lenz's law*. We can determine this polarity easily by the example illustrated by the loop in Fig. 3.12. A resistor of negligible length has been inserted between the two ends of the loop and the voltage drop V_{LOOP} is measured across it. When there is an incremental increase in the upward directed magnetic flux through the loop, $\Delta\Psi_m$, an incremental current through the loop, ΔI, must be in the CW direction to produce an incremental magnetic field and flux, $-\Delta\Psi_m$, which is directed downward in opposition to the original change in flux. This current produces a voltage drop across the resistor that is positive on the top and negative on the bottom. If we now take the limit as the resistor approaches an open circuit, we have the voltage with the correct polarity as defined by Eq. (3.23).

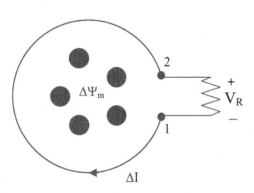

FIGURE 3.12: Illustration of Lenz's law.

Even though it is a PEC loop (with no voltage drop along the PEC), there is a voltage drop across its ends induced by the changing magnetic field that it encircles. Moreover, note that the current flows from the more negative end of the loop to the more positive end of the loop. This is definitely not the behavior of current in a resistive circuit; in fact, this behavior is that of an ideal voltage source. That is just what it is! The voltage defined by Eq. (3.23) is the open-circuit or Thévenin equivalent source. The changing magnetic flux activates the voltage source. To determine the Thévenin equivalent impedance, the source is deactivated by eliminating the changing magnetic field. This leaves only the loop of wire that has zero resistance. The PEC loop in a

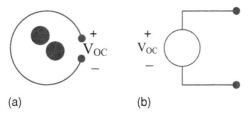

(a)   (b)

FIGURE 3.13: Thévenin equivalent circuit for a PEC loop in a magnetic field: (a) physical configuration and (b) equivalent circuit.

changing magnetic field acts as an ideal voltage source with zero resistance. The loop and its Thévenin equivalent circuit are shown in Fig. 3.13.

Expressing the flux enclosed by the PEC loop in terms of the magnetic flux density, we can write Faraday's law as

$$V_{\text{LOOP}} = -\frac{d\Psi_m}{dt} = -\frac{d}{dt}\left(\iint_S \mathbf{B} \cdot \mathbf{ds}\right) \qquad (3.24)$$

where S is any surface for which the PEC loop is the perimeter. Lenz's law establishes the polarity of the induced voltage at the ends of the loop. For **ds** directed into the plane of the paper in Fig. 3.12 as required by the direction of integration along the loop from point 1 to point 2 and **B** in the out of the plane of the paper, the flux integral is negative. In addition, if **B** is increasing, then the derivative is positive and the LHS is positive indicating $V_2 - V_1 > 0$. Application of Lenz's law also shows that $V_R = V_2 - V_1 > 0$.

Faraday's law describes the induced voltage in a loop without any restrictions upon the source of the magnetic flux. In applications where the source is separated from the coil, the induced voltage can indicate the presence of a magnetic field or it can serve to "couple" the induced voltage to the current that establishes the flux. In inductors, the voltage is developed across the loops of the coil in which the excitation current flows. The behavior of an inductor is the topic of the next section.

Example 3.7-1.   The magnetic field, which is defined as upward positive as shown in Fig. 3.14(a), is uniform throughout the PEC wire loop of 1 m radius. The magnetic field varies with time as shown in Fig. 3.14(b). Calculate the voltage observed by the voltmeter. From Eq. (3.2), the magnetic flux upward through the loop is given by $\Psi_m = \iint_S \mathbf{B} \cdot \mathbf{ds} = B_{\text{UP}}\pi(1)^2 = \pi B_{\text{UP}}$. The flux has the same variation with time as the flux density. Replace the voltmeter with a resistor and apply Lenz's law. An increasing upward flux requires a downward flux to oppose it; by the RH rule this requires a CW current in the loop. The

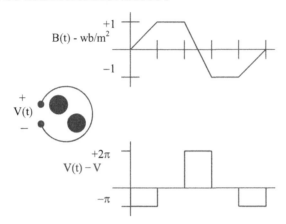

FIGURE 3.14: Induced voltage: (a) circuit geometry, (b) $B(t)$, and (c) $V(t)$.

lower end of the resistor (the negative terminal of the voltmeter) will be more positive. Consequently, the observed voltage will be negative for this case. Similarly, the observed voltage will be positive for a decreasing upward flux.

Since an induced voltage exists only when the flux is changing, there will be a nonzero, observed voltage only in the time intervals of (0,1), (2,3), and (4,5). During the interval $0 \le t \le 1$, $d\Psi_m/dt = \pi$ so the observed voltage will be $-\pi V$; during the interval $2 \le t \le 3$, $d\Psi_m/dt = -2\pi$ so the observed voltage will be $2\pi V$; during the interval $4 \le t \le 5$, $d\Psi_m/dt = \pi$ so the observed voltage will be $-\pi V$. The observed voltage is zero for all other times since the flux is unchanging during these times. The observed voltage $V(t)$ is shown in Fig. 3.14(c).

Example 3.7-2.   Calculate the observed voltage of Example 3.7-1 if the magnetic field varies as $B_{UP} = \sin 377t$ wb/m². As before the flux through the loop is $\psi_m = \pi B_{UP} = \pi \sin 377t$, $d\psi_m/dt = 377\pi \cos 377t$ and the observed voltage is $V(t) = -377\pi \cos 377t$ V. The continuous variations of the sinusoidal magnetic field produce a continuous induced voltage. However, the voltage is delayed in phase by 90° from the magnetic field. In addition, the voltage is proportional to the angular frequency of the magnetic field, in this case 377 rad/s; the higher the frequency of the magnetic field, the larger the induced voltage.

3.8   SELF INDUCTANCE

The concepts of the preceding sections provide us the background by which we can calculate the inductance of a coil of wire wound on a magnetic core. Consider the inductor shown in

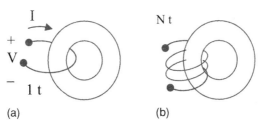

FIGURE 3.15: Magnetic core inductors: (a) one-turn inductor and (b) N-turn inductor.

Fig. 3.15. According to Ampere's law, the current flowing in the PEC coil wound on the magnetic core establishes a magnetic field within the core given by $H = NI/2\pi\rho$. The permeability of the core relates the magnetic field to the magnetic flux density by $B = \mu H$. The flux in the core is given as

$$\Psi_m = \iint_S \mathbf{B} \cdot \mathbf{ds} = \int_{z=0}^{t} \int_{\rho=a}^{b} \frac{\mu NI}{2\pi\rho} d\rho dz$$

$$= \frac{\mu NI}{2\pi} t \ln\left(\frac{b}{a}\right). \tag{3.25}$$

Consequently, the flux is proportional to the current. For DC currents, there is no voltage drop across the coil since there is neither a resistive voltage drop nor an induced voltage. However, if the current is time varying, it produces a time-varying magnetic flux which, according to Faraday's law, induces a voltage drop across each loop of the coil.

Before proceeding further, let's make a convenient approximation to Eq. (3.25). With the geometry of the toroid defined in terms of the mean or average radial distance, $\Delta = b - a$, and the average radius, $\rho_{\mathrm{MEAN}} = (a + b)/2$, the logarithm term of Eq. (3.25) is expressed as

$$\ln\left(\frac{b}{a}\right) = \ln\left(\frac{\rho_{\mathrm{MEAN}} + \Delta/2}{\rho_{\mathrm{MEAN}} + \Delta/2}\right) = \ln\left(\frac{1 + \frac{\Delta/2}{\rho_{\mathrm{MEAN}}}}{1 - \frac{\Delta/2}{\rho_{\mathrm{MEAN}}}}\right). \tag{3.26}$$

For the situation where the inner and outer radii, a and b, are nearly equal, $\Delta/\rho_{\mathrm{MEAN}} \ll 1$ and Eq. (3.26) simplifies as

$$\ln\left(\frac{1 + \frac{\Delta/2}{\rho_{\mathrm{MEAN}}}}{1 - \frac{\Delta/2}{\rho_{\mathrm{MEAN}}}}\right) \approx \ln\left(\left(1 + \frac{\Delta/2}{\rho_{\mathrm{MEAN}}}\right)\left(1 + \frac{\Delta/2}{\rho_{\mathrm{MEAN}}} + \ldots\right)\right)$$

$$= \ln\left(1 + \frac{\Delta}{\rho_{\mathrm{MEAN}}} + \ldots\right) \approx \frac{\Delta}{\rho_{\mathrm{MEAN}}}. \tag{3.27}$$

This approximation is accurate to within 10% for $\Delta \leq \rho_{\text{MEAN}}$, so it is actually valid for the vast majority of cases we will consider. Moreover, for incremental inductances (which we will define shortly), it is exactly correct, not just an approximation. With this substitution, Eq. (3.25) becomes

$$\Psi_m = \frac{\mu N I}{2\pi} t \ln\left(\frac{b}{a}\right) \approx \frac{\mu N I}{2\pi} t \frac{\Delta}{\rho_{\text{MEAN}}}$$

$$= \frac{\mu N I t \Delta}{2\pi \rho_{\text{MEAN}}} = \frac{\mu N I A}{l_{\text{MEAN}}}. \tag{3.28}$$

where $t\Delta = A$ is the cross-sectional area of the core and $2\pi \rho_{\text{MEAN}} = l_{\text{MEAN}}$ is the mean length of the flux path within the core. Not only does this form simplify the expression for flux, it allows a convenient geometric interpretation. This approximation assumes that the flux is uniform throughout the cross section of the core and that all of the flux paths are of the same length, l_{MEAN}. This behavior is analogous to the flux in axial resistors and parallel plate capacitors. Furthermore, it enables us to use many of the solution techniques we have already learned. Hereafter, we will use l to represent l_{MEAN}.

Now, back to using Faraday's law to calculate inductance. Let's begin with a single-turn coil, i.e., $N = 1$, for which the voltage is expressed as

$$V_{\text{LOOP}} = -\frac{d\Psi_m}{dt} = \frac{d}{dt}\left(\frac{\mu A I}{\ell}\right) = \left(\frac{\mu A}{\ell}\right)\frac{dI}{dt}. \tag{3.29}$$

The polarity of the voltage is determined by Lenz's law and is indicated in Fig. 3.15. Noting that the voltage drop is in the direction of the current flow, we rewrite Eq. (3.29) in the form used in circuit theory for inductors as

$$V_L = -V_{\text{LOOP}} = \frac{d\Psi_m}{dt} = \frac{d}{dt}\left(\frac{\mu A I}{\ell}\right) = \left(\frac{\mu A}{\ell}\right)\frac{dI}{dt}. \tag{3.30}$$

A comparison of Eqs. (3.1) and (3.30) leads to the an expression for the self-inductance of the one-turn loop as

$$L_{1\text{TURN}} = \frac{\mu A}{l}. \tag{3.31}$$

This shows the same geometric and material dependence that observe for expressing conductance and capacitance—the ratio of area in which a uniform flux exists to the length of the flux path times the constitutive property of the material that relates flux density to field intensity.

To generalize this result to an inductor with N turns or loops, we note that for linear cores the magnetic flux with N turns is N times the flux with one turn. This results in an N-fold increase in the induced voltage across each turn. Each turn is connected in series with adjacent turns; the total voltage across the inductor is N times the voltage induced in a single turn, see Fig. 3.15(b). Consequently, the inductance of an N-turn inductor is N^2 times the self-inductance of a one-turn inductor or

$$L = \frac{N^2 \mu A}{l} = N^2 L_{1T}. \tag{3.32}$$

Quite obviously, an increase in N causes a dramatic increase in the self-inductance.

The term *self-inductance* is used to describe the property by which a time-varying current in a coil induces a voltage in that coil. This is in contrast to the phenomenon of a current in one coil inducing a voltage in a second coil, known as *mutual inductance*. In keeping with the terminology of circuits, we will also use the term *inductance* to refer to self-inductance throughout this text.

Each term of Eq. (3.32) offers some insight into the concept of inductance. An increase in N increases the magnetic field intensity and the total magnetic flux that in turn increases the voltage induced in each turn of the inductor. An increase in N also increases the number of single-turn voltages that are connected in series and summed to give the inductor voltage. An increase in μ increases the magnetic flux density for a given magnetic field intensity and also increases the inductor voltage. An increase in A increases area within which the magnetic flux density is contained with a corresponding increase in the total magnetic flux. A decrease in the flux path l increases the magnetic field intensity for a given current. On the other hand, simultaneously increasing the inductance with all of these factors at the same time is not always possible. For example, as the length is decreased, the opening in the center of the core is decreased and the number of turns that can physically fit through the center of the core is decreased. Finally, we should note that as the number of turns is increased or the diameter of the wire is decreased in order that more turns can fit through the center, the resistance of an actual wire increases. At some point, the approximation of PEC wires will become invalid and an added resistance term must be included in the model for an inductor.

An alternate definition of inductance is based on Eq. (3.28), $\Psi_m = \frac{\mu N I A}{l}$. In an N-turn coil, Ψ_m is linked by each of the N turns for a total flux of $N\Psi_m$. $\lambda = N\Psi_m$ is also known as *flux linkages*. This leads to a definition of inductance as the ratio of the flux linkages of the coil to the current flowing in the coil as

$$L = \frac{\lambda}{I_L} = \frac{N\Psi_m}{I_L} = \frac{N^2 \mu A I_L}{l} \frac{1}{I_L} = \frac{N^2 \mu A}{l}. \tag{3.33}$$

This is in agreement with Eq. (3.32) and is analogous to the definition of capacitance based upon the charge/volt where charge is equivalent to the electric flux.

Finally, the model of uniform flux density and equal length of all flux paths is convenient and reasonably accurate for most engineering estimates of inductance. A further generalization is to consider these concepts as valid for any cross-sectional shape. This is especially helpful for those shapes for which computations are difficult.

Example 3.8-1.   Calculate the inductance of a 100-turn coil wound on a toroidal core with inner radius of 3 cm and an outer radius of 4 cm, a thickness of 2 cm, and a relative permeability of 1000. Application of Eq. (3.32) gives

$$L = \frac{N^2 \mu A}{l} = \frac{(100)^2 10^3 4\pi \times 10^{-7}(0.01)(0.02)}{2\pi(0.035)} = 11.4 \text{ mH.}$$

3.9   MAGNETOMOTIVE FORCE

The behavior of resistors, capacitors, and inductors is proportional to the amount of flux within the element. In addition, resistance and capacitance are inversely proportional to the voltage drop or potential difference across the device. The similarity between the expression for inductance and those for conductance and capacitance suggests that we might extend this concept by defining the magnetic equivalent of voltage drop as

$$V_{\text{mAB}} = V_{\text{mA}} - V_{\text{mB}} = -\int_B^A \mathbf{H} \cdot \mathbf{dl}, \qquad (3.34)$$

which is known as *magnetomotive force* or *mmf* with units of Ampere-turns. When the integration path is a single closed loop around the core, Ampere's law predicts the *mmf* drop as

$$V_{\text{mCORE}} = \oint_L \mathbf{H} \cdot \mathbf{dl} = \int_L \mathbf{H} \cdot \mathbf{dl} + \int_\delta \mathbf{H} \cdot \mathbf{dl}$$

$$= I_{\text{ENC}} = NI \qquad (3.35)$$

where N is the number of turns of the coil and I is the current in the coil as shown in Figure 3.16. A further extension of the analogy suggests that the mmf drop is in the direction of the flux. Since the integral around a closed path is nonzero, the magnetic field is not conservative. Instead, the total mmf drop is equal to the NI source of excitation of the magnetic field.

A mathematical artifice allows us to make the magnetic field within the core appear "conservative." If we restrict the path of the integral of Eq. (3.35) by erecting

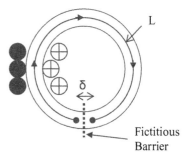

FIGURE 3.16:   Path constraint for conservative V_m.

a fictitious barrier at any cross section within the core, see Fig. 3.16, then there can never be a completely closed path. A path L^- that starts on one side of the barrier and progresses through the core to the other side of the barrier can be made arbitrarily close to L, but never quite closed. In the limit as $\delta \to 0$, the mmf drop calculated via the integral on path L^- still approaches and equals the value on path L. This artificial limitation makes $\mathbf{H}$ appear conservative within the core and, when limited to this region only, it behaves much as $\mathbf{E}$ within resistors and capacitors. This means that a differential form of Eq. (3.34) is valid as

$$\mathbf{H} = -\nabla V_m, \tag{3.36}$$

which can be combined with the continuity of magnetic flux, $\nabla \cdot \mathbf{B} = 0$, and the constitutive relation, $\mathbf{B} = \mu \mathbf{H}$, to give

$$\nabla \cdot \mathbf{B} = \nabla \cdot (\mu \mathbf{H}) = \nabla \mu \cdot \mathbf{H} + \mu \nabla \cdot \mathbf{H}. \tag{3.37}$$

For uniform magnetic material, $\nabla \mu = 0$ so that Eq. (3.37) becomes

$$\nabla \cdot \mathbf{B} = \nabla \cdot (\mu \mathbf{H}) = \mu \nabla \cdot \mathbf{H} = -\mu \nabla^2 V_m = 0. \tag{3.38}$$

The mmf satisfies Laplace's equation, too! Consequently, the magnetic flux and mmf equipotentials are perpendicular to each other. All of the properties of the resistor and capacitor solutions hold for inductors. Of course, we can use the same solution methods—analytic, curvilinear squares, iteration, linear equations, circuit analogs, and energy methods. The main difference is that magnetic flux tubes do not begin or end on electrodes, but are continuous, enclosed by the source of magnetic field—the current-carrying coil.

Finally, an incremental magnetic element analogous to the resistor and capacitor counterparts would be useful. It would be nice if it behaved as an inductor so that we could combine incremental elements as in circuits. However, this is not possible due to several aspects of the inductor—the absence of electrodes, the magnetic flux through the coil, and the current perpendicular to the flux. However, by defining an incremental reluctor (where reluctance is the reciprocal of inductance), we can apply the same circuit combination rules that we used with conductors and capacitors. In the next chapter, we will further utilize this concept with magnetic devices. The incremental reluctance for a coil of one turn is defined as

$$\Delta \mathcal{R}_{1T} = \frac{\Delta V_m}{\Delta \Psi_m} = \frac{\Delta l}{\mu \Delta a} \; [\mathrm{H}^{-1}]. \tag{3.39}$$

It represents the "resistance" of a magnetic core to accept magnetic flux. The incremental reluctor has the same curvilinear nature of incremental resistors and capacitors. All surfaces are perpendicular to each other, see Fig. 3.17. A uniform mmf drop, ΔV_m, occurs across the length of the incremental reluctor, Δl. A uniform magnetic flux, $\Delta\Psi_m$, penetrates the entire cross-sectional area, Δa. This is the incremental reluctance due current flowing through a one-turn coil. The longer the core the greater the reluctance; the greater the cross-sectional area the less the reluctance.

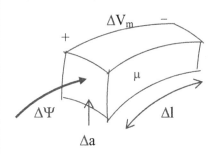

FIGURE 3.17: Incremental reluctor.

The advantage of reluctance is that incremental reluctances combine in the same fashion as resistors in circuits, i.e., series reluctances combine as series resistance and parallel reluctances combine as parallel resistance. After obtaining the total reluctance from the series–parallel combinations, we calculate inductance as the reciprocal of reluctance according to

$$L_{1T} = \frac{1}{\mathcal{R}_{1T}} \qquad (3.40)$$

The reluctance of a finite structure is the series–parallel combination of incremental reluctors where reluctances combine as resistances. The reluctance of series mmf connections within a flux tube is given by

$$\Delta\mathcal{R}_{\text{SERIES}} = \Delta\mathcal{R}_1 + \Delta\mathcal{R}_2 = \frac{\Delta\ell_1}{\mu\Delta a_1} + \frac{\Delta\ell_2}{\mu\Delta a_2} = \frac{\Delta\ell_1 + \Delta\ell_2}{\mu\Delta a}, \qquad (3.41)$$

see Fig. 3.18(a). The inductance of this series connection of reluctors is then calculated as

$$\Delta L_{\text{SERIES}} = \frac{1}{\Delta\mathcal{R}_{\text{SERIES}}} = \frac{\mu\Delta a}{\Delta l_1 + \Delta l_2}.$$

The reluctance of parallel flux tubes is given by

$$\Delta\mathcal{R}_{\text{PARALLEL}} = \left(\frac{1}{\Delta\mathcal{R}_1} + \frac{1}{\Delta\mathcal{R}_2}\right)^{-1} = \left(\frac{\mu\Delta a_1}{\Delta l_1} + \frac{\mu\Delta a_2}{\Delta l_2}\right)^{-1}$$

$$= \left(\frac{\mu(\Delta a_1 + \Delta a_2)}{\Delta l}\right)^{-1} = \frac{\Delta l}{\mu(\Delta a_1 + \Delta a_2)}, \qquad (3.42)$$

see Fig. 3.18(b). The inductance of this parallel connection of reluctors is then calculated as

$$\Delta L_{\text{PARALLEL}} = \frac{1}{\Delta\mathcal{R}_{\text{PARALLEL}}} = \frac{\mu(\Delta a_1 + \Delta a_2)}{\Delta l}.$$

A Word of Warning. The incremental inductances used here do NOT combine in the manner as circuit inductances since the mmf/flux ratio is not a circuit impedance. Instead, it is a sort

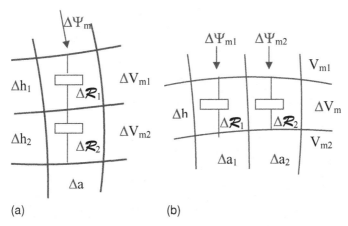

FIGURE 3.18: Series–parallel incremental reluctors: (a) series and (b) parallel.

of "magnetic impedance." But, when combined in manner of Eqs. (3.43) and (3.44), they will give the correct terminal value for an inductive structure.

3.10 MAGNETIC ENERGY STORAGE

The instantaneous power flow into an inductor is calculated from circuit concepts as

$$P_L = V_L I_L = L I_L \frac{d I_L}{dt} = L \frac{d}{dt} \left(\frac{I_L^2}{2} \right) \tag{3.43}$$

where I_L is the current flow in the direction of the voltage drop V_L. The power delivered to the inductor is positive when I_L^2 is increasing; it is negative when I_L^2 is decreasing; it is zero when the voltage is unchanging. With power as the time derivative of energy, we obtain

$$W_L = \frac{L I_L^2}{2}. \tag{3.44}$$

No energy is stored when the current is zero; the sign of energy is always positive regardless of the polarity of the applied current. While the current increases, power is delivered to the inductor as the energy stored in the inductor is increasing. While the current decreases, the inductor delivers power back to the circuit as the energy stored in the inductor decreases. While the current is constant, no power flow occurs and the energy stored is unchanged for ideal magnetic material.

Since the current is related to the magnetic field, the energy stored in the inductor can be expressed in terms of the fields. For a one-turn inductor, Eq. (3.44) becomes

$$
W_{L1T} = \frac{L I_L^2}{2} = \frac{1}{2} \left(\frac{\mu \iint\limits_S \mathbf{H} \cdot \mathbf{ds}}{\int\limits_L \mathbf{H} \cdot \mathbf{dl}} \right) \left(\int\limits_L \mathbf{H} \cdot \mathbf{dl} \right)^2
$$

$$
= \frac{1}{2} \left(\iint\limits_S \mathbf{H} \cdot \mathbf{ds} \right) \left(\int\limits_L \mathbf{H} \cdot \mathbf{dl} \right). \tag{3.45}
$$

An energy related expression of inductance is obtained by considering the incremental inductor of Fig. 3.17. An expression for the incremental energy stored in the incremental inductor is obtained from Eq. (3.46) as

$$
\Delta W_L = \frac{\Delta L (\Delta I_L)^2}{2} = \frac{(\Delta I_L)^2}{2 \Delta l}
$$

$$
= \frac{\mu \Delta a}{2 \Delta l} (|\mathbf{H}| \Delta l)^2 = \frac{\mu |\mathbf{H}|^2}{2} \Delta \mathbb{V} \tag{3.46}
$$

where $\Delta \mathbb{V}$ is the volume of the incremental inductor. In a fashion similar to capacitors, *magnetic energy density* is defined as

$$
w_m = \frac{\Delta W_L}{\Delta \mathbb{V}} = \frac{\mu |\mathbf{H}|^2}{2} = \frac{\mu \mathbf{H} \cdot \mathbf{H}}{2} = \frac{\mathbf{H} \cdot \mathbf{B}}{2}. \tag{3.47}
$$

This correctly represents the magnetic energy density for all magnetic fields. The total magnetic energy within a finite volume is expressed similarly to the electric energy storage as

$$
W_m = \sum_{\text{VOLUME}} \Delta W_m = \sum_{i=1}^{N} w_{mi} \Delta v_i = \sum_{i=1}^{N} \frac{\mu_i |\mathbf{H}_i|^2}{2} \Delta v_i \tag{3.48}
$$

for numeric calculations and

$$
W_m = \iiint\limits_{\mathbb{V}} \frac{\mu |\mathbf{H}|^2}{2} dv \tag{3.49}
$$

for analytic calculations.

Equations (3.11) and (3.49) are combined to express the definition of inductance in terms of stored magnetic energy as

$$L = \frac{2W_m}{I_L^2} = \frac{2 \iiint_V \frac{\mu |\mathbf{H}|^2}{2} dv}{\frac{1}{N^2}\left(\int_L \mathbf{H} \cdot \mathbf{dl}\right)^2} = \frac{N^2 \iiint_V \mu |\mathbf{H}|^2 dv}{\left(\int_L \mathbf{H} \cdot \mathbf{dl}\right)^2}. \tag{3.50}$$

This form suggests that those regions of space where the magnetic field is great can be thought of as inductive. Often this region can be represented by a lumped inductor. This form is preferred when using optimization methods in which the principle of energy minimization is utilized.

3.11 INDUCTANCE CALCULATIONS
Several calculation methods of inductance are shown in the examples below.

Example 3.11-1. Calculate the inductance of the equilateral, triangular shaped core shown in Fig. 3.18 by the method of curvilinear squares. The core is 1 cm thick, 0.5 cm high, and has $\mu_R = 1000$. A coil of 100 turns is wound on it.

Due to the symmetry, curvilinear squares are required for only one-sixth of the core, see Fig. 3.19. As with resistors and capacitors, the key is to make the magnetic equi-mmf (dashed lines), V_m, and flux lines (solid lines), Ψ_m, perpendicular. Curvilinear reluctances add in the manner of resistances so the total one-turn inductance is calculated as $L_{1T} = 1/R_{1T} = \mu t[(3 + 3||5)^{-1} + 1/5 + 1/5]/6 = (0.61)4\pi \times 10^{-7}(1000)(0.01)/6 = 1.27\,\mu\text{H}$. Therefore, the total inductance is $L = N^2 L_{1T} = 12.7$ mH.

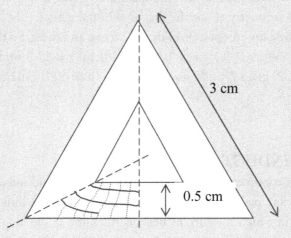

FIGURE 3.19: Equilateral triangular core.

Example 3.11-2.   Calculate the inductance of Example 3.11-1 by using the mean flux path length. Recall that this is the same as assuming that the flux density is constant over each cross section. From the geometry, (dust off your trig skills!) the mean path length is calculated as $l = 3(0.0257) = 0.077$ m. The inductance is given by $L = (100)^2 4\pi \times 10^{-4}(0.01)(0.005)/0.077 = 8.2$ mH a reasonable approximation of the solution of Example 3.11-1.

Example 3.11-3.   Calculate the inductance of Example 3.11-1 by the energy method. Assume uniform flux density. The volume of the core is calculated as $V = 1_{MEAN}A = 0.077 * 0.005 * 0.01 = 3.85 \times 10^{-6}$ m^3. From Equation **Error! Reference source not found.**, the inductance is calculated as

$$L = \frac{2W_m}{I_L^2} = \frac{N^2 \iiint_V \mu |\mathbf{H}|^2 dv}{\left(\int_L \mathbf{H} \cdot d\ell\right)^2} \approx \frac{\mu(NH)^2 V}{(H\ell)^2} = \frac{\mu N^2 V}{(\ell)^2} = \frac{4\pi \, 10^{-4}(100)^2 3.85 \times 10^{-6}}{(0.077)^2} = 8.2 \text{ mH.}$$

Example 3.11-4.   Calculate the inductance of the rectangular core in Fig. 3.20 with $\mu_R = 1000$, a thickness of 1 cm, and a 100-turn coil using magnetic potential and spreadsheet techniques. The mean path length is 14 cm. For convenience we can perform the calculations with only the upper-left quadrant. If we assume that the total mmf drop for the entire core is 1 At, then the mmf drop across the quadrant is 0.25 At. The magnetic potential displays the expected symmetry about the diagonal in the corner. The Ψ_m / V_m, Eq. (3.53), calculation is normalized for unit permeability, thickness, and turns. So the inductance of the core is $L = N^2 \mu t(0.138) = (100)^2 4\pi \times 10^{-4}(0.01)(0.138) = 17.3$ mH. The uniform flux solution with Eq. (3.32) gives $L = N^2 \mu A/l = (100)^2 4\pi \times 10^{-4}(0.01)(0.015)/0.14 = 13.6$ mH.

3.12   MUTUAL INDUCTANCE

A changing current in a loop establishes a changing magnetic field intensity and flux density. When this changing flux passes through a closed loop of wire it induces a voltage in that loop. When the loop in which the current flows is the same as the loop in which the voltage is induced, the parameter of *self-inductance* is used to describe the relationship between the

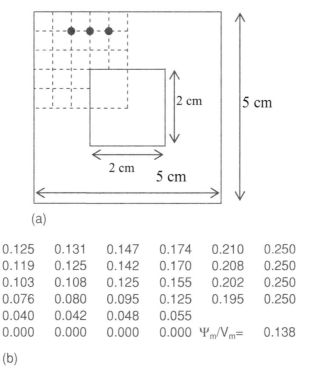

(a)

0.125	0.131	0.147	0.174	0.210	0.250
0.119	0.125	0.142	0.170	0.208	0.250
0.103	0.108	0.125	0.155	0.202	0.250
0.076	0.080	0.095	0.125	0.195	0.250
0.040	0.042	0.048	0.055		
0.000	0.000	0.000	0.000	$\Psi_m/V_m=$	0.138

(b)

FIGURE 3.20: Inductance calculation by spreadsheet: (a) geometry and (b) spreadsheet calculations.

changing current and the induced voltage. On the other hand, the phenomenon of a changing magnetic flux inducing a voltage in a coil is more general. The changing flux can be linked by a second coil in which a voltage is induced. The relationship between a changing current in one coil and an induced voltage in a second coil is known as *mutual inductance*. From circuits we know this relationship as

$$V_2 = M_{21}\frac{dI_1}{dt} \tag{3.51}$$

where M_{21} is the mutual inductance that defines the voltage induced in the second coil due a changing current in the first coil. Mutual inductance has units of Henries. The two coils are "linked" by the magnetic flux between them. The process is *bilateral*. That is, if a changing current in coil one induces a voltage in coil two, then the same current in coil two will induce the same voltage in coil one. This is expressed as $M_{12} = M_{21} = M$. There may be many coils linked by the same flux with a mutual inductance term for each linkage. Calculations of mutual inductance are similar to those of self-inductance except that the flux is generated by a current outside of the coil in which the voltage is induced.

The basic definition of an incremental mutual inductance for both coils consisting of a single turn follows from Eq. (3.38) as

$$\Delta M_{1T} = \frac{\Delta \Psi_{m1}}{\Delta V_{m2}} = \frac{\mu \Delta a_2}{\Delta l_1}. \qquad (3.52)$$

Current flow in an N_1-turn coil increases the flux by N_1 times; the voltage induced in an N_2-turn coil is increased by a N_2 over that of a single turn. The incremental mutual inductance is generalized to

$$\Delta M = \frac{\Delta \Psi_{m1}}{\Delta V_{m2}} = \frac{\mu N_1 N_2 \Delta a_2}{\Delta l_1}. \qquad (3.53)$$

The mutual inductance of two coils on a finite core follows from Eq. (3.32) as

$$M = \frac{\mu N_1 N_2 A}{l}. \qquad (3.54)$$

In the manner of Eq. (3.33), mutual inductance can be defined as the ratio of the flux linkages of the coupled coil to the current that establishes the flux as

$$M = \frac{N_2 \Psi_{m21}}{I_1} = \frac{\mu N_1 N_2 A}{l}. \qquad (3.55)$$

The voltage induced in the second coil acts as a controlled voltage source with zero Thévenin internal impedance and with an amplitude that depends upon the time rate of change of the current in the other coil. When the current is unchanging, there is no voltage induced in the second coil. In a manner similar to the blocking of DC currents by capacitors, DC currents do not induce a voltage in the second coil.

An alternate parameter is often used to describe the coupling between two coils. The *coupling coefficient*, k, is the ratio of the flux that links the second coil, Ψ_{m21}, to the flux that links the current-carrying coil, Ψ_{m1}, i.e.,

$$k = \frac{\Psi_{m21}}{\Psi_{m1}}. \qquad (3.56)$$

When $k = 0$, no flux due to the first coil links the second coil; when $k = 1$, all the flux due to the first coil links the second. The cases we have examined so far has only one flux path (with no leakage flux) so that the flux in the two coils is the same and $k = 1$. This is because there is no leakage from the core and the flux is the same in all parts of the core. As you recall from circuits, k can be expressed in terms of the self- and mutual inductances of the two coils as

$$k = \frac{M}{\sqrt{L_1 L_2}} \qquad (3.57)$$

We shall return to consideration of mutually coupled coils in the next chapter when we look at transformers.

The polarity of the induced voltage is calculated by Lenz's law as with self-inductance—the induced voltage in the second coil has a polarity which tends to establish a current that opposes the original change in flux. The direction of current in the first coil is required in this determination in order to determine the original flux direction.

Example 3.12-1. Calculate the mutual inductance between an additional 200-turn coil and the existing 100-turn coil of Example 3.8-1. From Eq. (3.54), the mutual inductance is calculated as $M = \frac{N_1 N_2 \mu A}{l} = \frac{100(200)10^3 4\pi \times 10^{-7}(0.01)(0.02)}{2\pi(0.035)} = 22.8 \text{ mH}.$

Example 3.12-2. Determine the polarity of the induced voltages, V_{IND1} and V_{IND2}, for the core shown in Fig. 3.21. The current is defined as into the upper terminal of the exciting coil on the left, establishing a CW flux in the core. To counter changes in flux due to an increasing I_1, a current flowing into the bottom terminal (and out of the top) of coil one (upper right) is needed. Therefore, the top terminal is positive for V_{IND1}. On the other hand, the second coil is wound oppositely so that the current must flow into the top terminal. Consequently, the bottom terminal is positive for V_{IND2}.

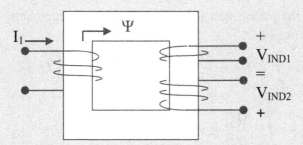

FIGURE 3.21: Polarity of induced voltage.

3.13 AMPERE'S LAW REVISITED AND THE CURL OPERATOR

The integral form of Ampere's law,

$$\oint_L \mathbf{H} \cdot \mathbf{dl} = I_{\text{ENC}}, \tag{3.58}$$

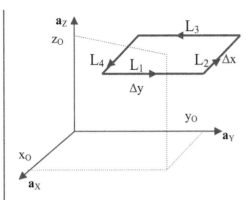

FIGURE 3.22: Incremental path for Ampere's law.

provides global information regarding the relationship between the magnetic field intensity and the current that establishes it. Localized behavior can be obtained by evaluating Eq. (3.6) as the path L shrinks to incremental size. This is not entirely new to us since we have used similar methods in obtaining $\nabla \cdot \mathbf{D}$ from the integral form of Gauss' law. To accomplish this, let's evaluate Eq. (3.6) along a closed rectangular path L_Z in the $z = z_o$ plane and centered at (x_o, y_o), see Fig. 3.22. The magnetic field intensity at the center of the path is expressed as $\mathbf{H} = H_{XO}\mathbf{a}_X + H_{YO}\mathbf{a}_Y + H_{ZO}\mathbf{a}_Z$.

The integral is expressed as the sum of integrals along the four segments

$$\oint_{L_Z} \mathbf{H} \cdot \mathbf{d\ell} = \int_{L_1} H_Y dy + \int_{L_2} H_X dx + \int_{L_3} H_Y dy + \int_{L_4} H_X dx. \tag{3.59}$$

Since the paths are only incrementally long, the fields on the paths 1 and 3 can be approximated by a Taylor series expansion as

$$H_Y = H_{YO} \pm \frac{\partial H_{YO}}{\partial x}\frac{\Delta x}{2} + \frac{\partial^2 H_{YO}}{\partial x^2}\frac{(\Delta x)^2}{8} + \cdots \tag{3.60}$$

where the $+$ sign is on path 1 and the $-$ sign on path 3. The integrals of paths 1 and 3 are combined and, since the paths are of incremental length, are approximated as the integrand times the path length to give

$$\int_{L_1} H_Y dy + \int_{L_3} H_Y dy = \int_{\Delta x} \left(H_{YO} + \frac{\partial H_{YO}}{\partial x}\frac{\Delta x}{2} + \cdots \right) dy - \int_{\Delta x} \left(H_{YO} - \frac{\partial H_{YO}}{\partial x}\frac{\Delta x}{2} + \cdots \right) dy$$

$$= \int_{y_o - \frac{\Delta y}{2}}^{y_o + \frac{\Delta y}{2}} \frac{\partial H_{YO}}{\partial y}\Delta x dy \approx \frac{\partial H_{YO}}{\partial x}\Delta x \Delta y. \tag{3.61}$$

In a similar manner, the other two segments of the path give

$$\int_{L_2} H_X dx + \int_{L_4} H_X dx \approx -\frac{\partial H_{XO}}{\partial y}\Delta x \Delta y. \tag{3.62}$$

These results substituted into the RHS of Eq. (3.6) give

$$I_{ENC} = \oint_{L_Z} \mathbf{H} \cdot \mathbf{d\ell} \approx \left(\frac{\partial H_{YO}}{\partial x} - \frac{\partial H_{XO}}{\partial y} \right) \Delta x \Delta y. \tag{3.63}$$

When only volume current densities are present, the RHS of Eq. (3.6) gives

$$I_{\text{ENC}} = \iint_S \mathbf{J} \cdot \mathbf{ds} = \int_{y=y_0-\frac{\Delta y}{2}}^{y_0+\frac{\Delta y}{2}} \int_{x=x_0-\frac{\Delta x}{2}}^{x_0+\frac{\Delta x}{2}} J_Z dx dy$$

$$\approx J_{ZO} \Delta x \Delta y \tag{3.64}$$

for the incremental surface of Fig. 3.22. Ampere's law requires that these last two expressions are equal as Δx, $\Delta y \to 0$, i.e.,

$$\frac{\partial H_{YO}}{\partial x} - \frac{\partial H_{XO}}{\partial y} = J_{ZO}. \tag{3.65}$$

In the limit this can be written as

$$\lim_{\Delta x, \Delta y \to 0} \frac{\oint_{L_Z} \mathbf{H} \cdot \mathbf{d\ell}}{\Delta x \Delta y} = \frac{\partial H_Y}{\partial x} - \frac{\partial H_X}{\partial y} = J_X. \tag{3.66}$$

As the surface becomes vanishingly small, the line integral of the magnetic field intensity per unit area in the $z = z_o$ plane is equal to the volume current flux density perpendicular to that plane. Moreover, it is equal to the differences in the rate of change of the x- and y-components of the field intensity. We have arrived at the desired relationship of the magnetic field at a point in terms of the volume current density. The line integral in Eq. (3.66) is known as the z-component of the *circulation* of the magnetic field intensity.

Similarly, this process can be applied to J_X and J_Y to obtain

$$\lim_{\Delta y, \Delta z \to 0} \frac{\oint_{L_X} \mathbf{H} \cdot \mathbf{d\ell}}{\Delta y \Delta z} = \frac{\partial H_Z}{\partial y} - \frac{\partial H_Y}{\partial z} = J_X \tag{3.67}$$

and

$$\lim_{\Delta x, \Delta z \to 0} \frac{\oint_{L_Y} \mathbf{H} \cdot \mathbf{d\ell}}{\Delta x \Delta z} = \frac{\partial H_X}{\partial z} - \frac{\partial H_Z}{\partial x} = J_Y. \tag{3.68}$$

The three components are combined as a vector

$$\left(\frac{\partial H_Z}{\partial y} - \frac{\partial H_Y}{\partial z} \right) \mathbf{a}_X + \left(\frac{\partial H_X}{\partial z} - \frac{\partial H_Z}{\partial x} \right) \mathbf{a}_Y + \left(\frac{\partial H_Y}{\partial x} - \frac{\partial H_X}{\partial y} \right) \mathbf{a}_Z = \mathbf{J} \tag{3.69}$$

to obtain a single expression that shows the local relationship between the components of $\mathbf{H}$ and of $\mathbf{J}$. The expression on the LHS of Eq. (3.69) is known commonly as the *curl of* $\mathbf{H}$. Its German equivalent is *rot of* $\mathbf{H}$ due to its connection with the rotation or circulation of $\mathbf{H}$.

A close examination of the form of Eq. (3.69) reveals that it can be expressed in terms of the nabla operator, ∇, used in the gradient and divergence. If the nabla operator is treated as a vector, Eq. (3.69) is written as

$$\nabla \times \mathbf{H} = \mathbf{J}. \qquad (3.70)$$

This is the *point form of Ampere's law* expressed in formal vector notation. The curl is written as

$$\nabla \times \mathbf{H} = \left(\frac{\partial H_Z}{\partial y} - \frac{\partial H_Y}{\partial z} \right) \mathbf{a}_X + \left(\frac{\partial H_X}{\partial z} - \frac{\partial H_Z}{\partial x} \right) \mathbf{a}_Y + \left(\frac{\partial H_Y}{\partial x} - \frac{\partial H_X}{\partial y} \right) \mathbf{a}_Z \qquad (3.71)$$

in Cartesian coordinates. Of course, the curl can be expressed in cylindrical and spherical coordinate systems as well, see Appendix C.

Each component of the curl of the magnetic field intensity is nonzero only where there is a perpendicular current flux density. This means that there must be a current enclosed by the vanishingly small path used to describe that component of the curl. Throughout those regions where there is no current density, the curl vanishes, i.e., $\nabla \times \mathbf{H} = \mathbf{J} = 0$. This point form of Ampere's law is not valid at filamentary or surface currents since the spatial derivatives of the curl operator become infinite there; the integral form of Ampere's law is valid everywhere.

One interpretation of the curl appeared in Professor Hugh Skilling's pioneering textbook, *Fundamentals of Electric Waves*, as an analog with fluid flow. He suggests that a "curl meter" would take the form of a very small paddle wheel on a shaft. If the field has a nonzero component of the curl, the paddle wheel will turn when it is aligned with this plane. The speed of the shaft rotation is a measure of the strength of that component of the curl; the direction of rotation determines the sign. The paddle wheel can be oriented in each of three mutually orthogonal planes to find three orthogonal components of the curl. The vector sum of the three components is the curl of the field.

Examples of a curl meter in various two-dimensional flux fields are shown in Fig. 3.23. The field lines of a point charge are directed radially outward as shown in the upper-left figure. No matter how one orients the paddle wheel, there will never be a net torque to cause it to rotate, therefore, the curl of this field is zero. Likewise, in the uniform velocity field (upper-right figure), there is no net torque on the curl meter. Rotation doesn't occur and the field has a zero curl. On the other hand, the nonuniform velocity fields of the lower-right and -left figures exert a net torque on the curl meter. The lower-left field has a negative curl; the lower-right has a positive curl. When the axis of the paddle wheel is aligned with the curl $\mathbf{v}$ it will rotate most rapidly; the vector sum of the components of the curl is aligned in this direction.

DC electric fields have no curl since the work integral about a closed path is zero,

$$\nabla \times \mathbf{E} = 0. \qquad (3.72)$$

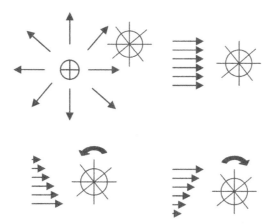

FIGURE 3.23: Curl meter measurement of curl.

Slowly varying electric fields have such a small value of curl that this can be accurately approximated as satisfying Eq. (3.72) as well. However, rapidly varying electric fields have a nonzero curl as we shall learn later. Let's defer the definition as to what we mean by rapidly varying until later as well. Fields that satisfy Eq. (3.72) are known as *irrotational* fields since their curl vanishes.

We have already learned that when the work integral of a field intensity vanishes for all possible paths in a region, the field is conservative throughout that region. But now we know that this means that the curl of the field intensity vanishes as well. From these results, we can infer that *whenever the curl of a field intensity is zero throughout a region, its work integral around all closed paths in that region vanishes and the field is conservative.* This property distinguishes a conservative field from a nonconservative field; it is a convenient test for the conservative nature of a field.

For any vanishingly small region of Fig. 3.24, the closed path integral can be approximated by the curl as

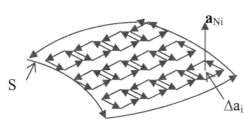

FIGURE 3.24: Demonstration of Stokes' theorem.

$$\oint_{\Delta L_i} \mathbf{H} \cdot \mathbf{dl} \approx (\nabla \times \mathbf{H})_i \, \Delta a_i = (\nabla \times \mathbf{H}) \cdot \mathbf{a}_{Ni} \Delta a_i$$

$$= (\nabla \times \mathbf{H}) \cdot \Delta \mathbf{s}_i \qquad (3.73)$$

where $(\nabla \times \mathbf{H})_i$ is the component of $\nabla \times \mathbf{H}$ in the direction of the unit normal $\mathbf{a}_{Ni}$ of the ith incremental area Δa_i.

As the work integral (the circulation) is computed around each incremental area, the results are

$$\oint_L \mathbf{H} \cdot \mathbf{dl} = \sum_{i=1}^{N} \oint_{\Delta L_i} \mathbf{H} \cdot \mathbf{dl} = \sum_{i=1}^{N} (\nabla \times \mathbf{H}) \cdot \Delta \mathbf{s}_i$$

$$= \iint_S (\nabla \times \mathbf{H}) \cdot \mathbf{ds} \qquad (3.74)$$

where the contributions of all of the interior paths cancel since the line integrals are in opposite directions for any two adjacent incremental areas.

Of course, the path direction and the surface normal satisfy the RH rule. This result shown in Eq. (3.74) is known as *Stokes' theorem* and provides us an alternative means of evaluating line integrals as long as the spatial derivative operations of the curl are finite. Note that this applies only to a closed line integral and results in an open surface integral.

Example 3.13-1.   Calculate the current density present for the magnetic field intensity $\mathbf{H} = I\rho \mathbf{a}_\phi / 2\pi a^2$. The curl in cylindrical coordinates is given as (see Appendix C)

$$\nabla \times \mathbf{H} = \left(\frac{1}{\rho} \frac{\partial H_z}{\partial \phi} - \frac{\partial H_\phi}{\partial z} \right) \mathbf{a}_\rho + \left(\frac{\partial H_\rho}{\partial z} - \frac{\partial H_z}{\partial \rho} \right) \mathbf{a}_\phi$$

$$+ \frac{1}{\rho} \left(\frac{\partial (\rho H_\phi)}{\partial \rho} - \frac{\partial H_\rho}{\partial \phi} \right) \mathbf{a}_z$$

so that for this magnetic field we have $\mathbf{J} = \nabla \times \mathbf{H} = \frac{1}{\rho} \frac{\partial (\rho H_\phi)}{\partial \rho} \mathbf{a}_z = \frac{I\mathbf{a}_z}{\pi a^2}$ A/m^2.

The current flux is uniform and perpendicular to the magnetic field intensity everywhere.

Example 3.13-2.   Is the field $\mathbf{G} = y\mathbf{a}_x + 2x\mathbf{a}_y$ a conservative field? Since the curl of a conservative field is identically zero and $\nabla \times \mathbf{G} = \left(\frac{\partial G_y}{\partial x} - \frac{\partial G_x}{\partial y} \right) \mathbf{a}_z = (2 - 1)\mathbf{a}_z \neq 0$, $\mathbf{G}$ is not conservative. However, if the x-component were multiplied by 2, then $\mathbf{G}$ would be conservative.

Example 3.13-3.   Calculate the current enclosed by the circular path of radius ρ for the magnetic field of Example 3.13-1 via Ampere's law. Repeat via Stokes' theorem. With Ampere's law, the enclosed current is $I_{\text{ENC}} = \int_{\phi=0}^{2\pi} \frac{I\rho}{2\pi a^2} \rho d\phi = \frac{\rho^2}{a^2} I$. With Stokes' theorem, the current enclosed is given by

$$I_{\text{ENC}} = \oint_L \mathbf{H} \cdot d\mathbf{l} = \iint_S (\nabla \times \mathbf{H}) \cdot d\mathbf{s}$$

$$= \int_{\rho'=0}^{\rho} \int_{\phi=0}^{2\pi} \left(\frac{1}{\rho'} \frac{\partial(\rho' H_\phi)}{\partial \rho'} \right) \rho' d\rho' d\phi$$

$$= \int_{\rho'=0}^{\rho} \int_{\phi=0}^{2\pi} \frac{I}{\pi a^2} \rho' d\rho' d\phi = \frac{\rho^2}{a^2} I.$$

As expected, the two methods give identical results.

3.14 ARBITRARY CURRENT DISTRIBUTIONS AND THE BIOT–SAVART LAW

Though it is global in nature, the integral form of Ampere's law is used to calculate directly the magnetic fields within a magnetic core. This is possible because the core behaves as an ideal flux guide with geometric symmetry. But, more general situations require a different approach. This is available through the differential form of the experimental Biot–Savart law as

$$d\mathbf{H}(\mathbf{r}) = \frac{d\mathbf{I}(\mathbf{r'}) \times \mathbf{a}_R}{4\pi R^2} \tag{3.75}$$

and is depicted in Fig. 3.25.

The differential magnetic field intensity of Biot–Savart shows a couple of similarities with the differential electric flux of Gauss, see Eq. (3.82). Both show an inverse square dependence upon the distance between the source and the field points. Both are proportional to the source. In contrast to the direction of the electric flux that is simply directed radially outward from the charge, the direction of the magnetic field is more complicated. The magnetic field is in the direction of the cross product of the current vector and the unit vector from the source to the field point. The magnetic field never points in the direction of the current, rather it is always perpendicular to it. Of course this is consistent with the RH rule that governs the global behavior of magnetic fields.

But before we get overly excited about the differential form, we must note that it is impossible to verify this form of the Biot–Savart law since differential current elements do not exist; we can observe complete loops of current only. The integral form has been shown to be

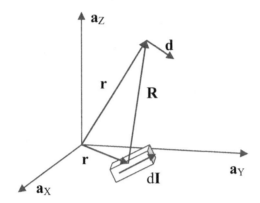

FIGURE 3.25: Application of the Biot–Savart law.

valid universally as

$$\mathbf{H}(\mathbf{r}) = \oint_{L_{\text{Current}}} \frac{d\mathbf{I}(\mathbf{r}') \times \mathbf{a}_R}{4\pi R^2} \qquad (3.76)$$

where the integration path is aligned with the closed circuit through which the current flows. In spite of this practical limitation to verification of the differential form of the Biot–Savart law, we will assume that we can apply it to all applications.

The form of the differential current depends upon the current distribution. For filamentary currents we have

$$d\mathbf{I}(\mathbf{r}') = I(\mathbf{r}')\mathbf{a}_I(\mathbf{r}')dl' \qquad (3.77)$$

where the current magnitude and its direction vary with the source coordinate $\mathbf{r}'$. For surface current density,

$$d\mathbf{I}(\mathbf{r}') = \mathbf{K}(\mathbf{r}')da'. \qquad (3.78)$$

For volume current density,

$$d\mathbf{I}(\mathbf{r}') = \mathbf{J}(\mathbf{r}')dv'. \qquad (3.79)$$

The integrals for these distributions are line, surface, and volume, respectively. The limits of integration are set to include all of the current.

This method is valid only for materials with homogeneous permeability. If there are permeability variations, alternative, more sophisticated methods must be used. These are deferred to a more advanced course.

The integrals have the same complexity as the integrals for arbitrary charge distribution—the integration requires the summation of vectors with varying amplitude and direction over the range of integration. As before, the simplest approach is to decompose the vectors into their Cartesian components that do not vary with position.

In addition to computation, a very useful feature of Eq. (3.75) is that it indicates the direction of the differential field intensity due to a differential current. This often reveals the addition or cancellation of the differential fields due to symmetrically located currents.

Example 3.14-1. Calculate the field of an infinitely long filament of uniform current I. From Fig. 3.26(a), we see that symmetrically located increments of current give rise to angularly-directed magnetic field intensity as indicated by the dots and crosses. In the vertical plane and above the current, the field components are out of the paper; below the current, they are into the paper. Moreover, they are of equal magnitude and have no angular dependence. Due to the infinite extent of the current, the fields have no variation with the z-position of the field point. For convenience, we pick the field point at $z = 0$. Moreover, it is convenient to align the current filament with the z-axis as shown in Fig. 3.26(b).

The field point is in the $z = 0$ plane and is defined as $\mathbf{r} = x\mathbf{a}_X + y\mathbf{a}_Y$. The source point is on the z-axis and is defined as $\mathbf{r}' = z'\mathbf{a}_Z$. Therefore, $\mathbf{R} = \mathbf{r} - \mathbf{r}' = x\mathbf{a}_X + y\mathbf{a}_Y - z'\mathbf{a}_Z$. The differential current element is defined as $d\mathbf{I} = I\mathbf{a}_Z dz'$. The differential flux density is expressed as

$$d\mathbf{H}(\mathbf{r}) = \frac{d\mathbf{I}(\mathbf{r}') \times \mathbf{a}_R}{4\pi R^2} = \frac{I\,dz'\mathbf{a}_Z \times (x\mathbf{a}_X + y\mathbf{a}_Y - z'\mathbf{a}_Z)}{4\pi[x^2 + z'^2]^{3/2}} = \frac{I\,dz'(x\mathbf{a}_Y - y\mathbf{a}_X)}{4\pi[x^2 + y^2 + z'^2]^{3/2}}.$$

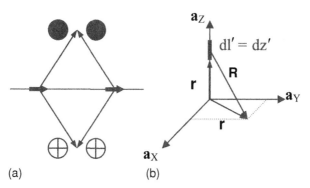

(a)　　　　　　　(b)

FIGURE 3.26: Infinitely long filament of uniform current: (a) differential fields and (b) geometry.

This leads to the total field at $(x, y, 0)$ as

$$
\mathbf{H}(\mathbf{r}) = \int_{z'=-\infty}^{\infty} \frac{I\,dz'(x\mathbf{a}_Y - y\mathbf{a}_X)}{4\pi\,[x^2 + y^2 + z'^2]^{3/2}} = \frac{I(x\mathbf{a}_Y - y\mathbf{a}_X)}{4\pi} \int_{z'=-\infty}^{\infty} \frac{dz'}{[x^2 + y^2 + z'^2]^{3/2}}
$$

$$
= \frac{I(x\mathbf{a}_Y - y\mathbf{a}_X)}{4\pi} \frac{2}{x^2 + y^2} = \frac{I}{2\pi\sqrt{x^2 + y^2}} \frac{(x\mathbf{a}_Y - y\mathbf{a}_X)}{\sqrt{x^2 + y^2}} = \frac{I\mathbf{a}_\phi}{2\pi\rho}.
$$

The field is totally in the angular direction; its magnitude depends only upon the radial distance from the current. Of course the result satisfies the RH rule. Note that the field and its derivatives become undefined at the origin due to the filamentary source.

Example 3.14-2.   Calculate the field along the axis of symmetry of a uniform current I flowing on a circular path of radius a. For convenience, the current is located in the $z = 0$ plane, centered on the origin, see Fig. 3.27. The field point is located on the z-axis so that $\mathbf{r} = z\mathbf{a}_Z$. The source point is on the loop of current, $\mathbf{r}' = a\mathbf{a}_{\rho'} = a(\cos\phi'\mathbf{a}_X + \sin\phi'\mathbf{a}_Y)$ where the unit vector $\mathbf{a}_{\rho'}$ that varies in direction with angular position on the current loop is replaced by its Cartesian coordinate equivalents. This leads to $\mathbf{R} = \mathbf{r} - \mathbf{r}' = z\mathbf{a}_Z - a(\cos\phi'\mathbf{a}_X + \sin\phi'\mathbf{a}_Y)$; $|\mathbf{R}| = [a^2 + z^2]^{1/2}$ has a constant length throughout the integration. The differential current element is expressed as $d\mathbf{I} = I\mathbf{a}_{\phi'}a\,d\phi' = Ia(-\sin\phi'\mathbf{a}_X + \cos\phi'\mathbf{a}_Y)d\phi'$. The differential field is given by

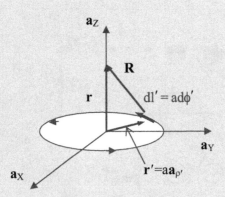

FIGURE 3.27: Loop of uniform current: (a) differential fields and (b) geometry.

$$dH(z, 0, 0) = \frac{dI(r') \times a_R}{4\pi R^2}$$

$$= \left\{ \begin{array}{c} \dfrac{Ia d\phi'(- \sin \phi' a_X + \cos \phi' a_Y)}{4\pi [a^2 + z^2]^{3/2}} \\ \times \dfrac{[za_Z - a(\cos \phi' a_X + \sin \phi' a_Y)]}{4\pi [a^2 + z^2]^{3/2}} \end{array} \right\}$$

$$= \frac{Ia d\phi'[z(\cos \phi' a_X + \sin \phi' a_Y) + a a_Z]}{4\pi [a^2 + z^2]^{3/2}}.$$

By inspection, the x- and y-components vanish since $\sin \phi$ and $\cos \phi$ integrate to zero over $0 - 2\pi$ range. Consequently, the magnetic field intensity is expressed as

$$H(z, 0, 0) = \int_{\phi'=0}^{2\pi} \frac{dI(r') \times a_R}{4\pi R^2}$$

$$= \int_{\phi'=0}^{2\pi} \frac{Ia^2 d\phi' a_Z}{4\pi [a^2 + z^2]^{3/2}} = \frac{Ia^2 a_Z}{2[a^2 + z^2]^{3/2}} \text{ A/m}.$$

The field along the axis is always in the z-direction and decreases as z^{-3} far from the current. At the center of the current loop, the field is simply $H(0, 0, 0) = Ia_Z/2a$ A/m. We purposely limited the calculations to points on the z-axis due to the complexity of analytic evaluations for off-axis field points. However, numeric solutions provide excellent results at all field points except on the current itself where the results become infinite as in Example 3.14-1.

3.15   FIELD INTENSITY VIA AMPERE'S LAW

As with Gauss' law, when there is sufficient symmetry, Ampere's law can be used to calculate the magnetic field intensity directly. Unfortunately, these idealized configurations are rarely realized in practice. Yet, they are so simple to compute that they often serve as prototypes with which we can approximate or compare actual current distribution of similar geometries. Several cases are considered in this section.

Consider the uniform, infinitely long, tube of current of radius a, shown in Fig. 3.28. This configuration closely approximates a thin wire within a circuit carrying a current I uniformly distributed as $|J| = I/\pi a^2$. Because of the symmetry of the geometry, the magnetic field intensity should be the same for any angle ϕ. Two symmetrically located differential elements of current produce fields for which the non-ϕ components cancel as shown in Fig. 3.28(a). Since the fields decrease with distance from the source, it seems likely that there will be a dependence of the field upon ρ, the radial distance from the current. As in Example 3.14-1, current elements

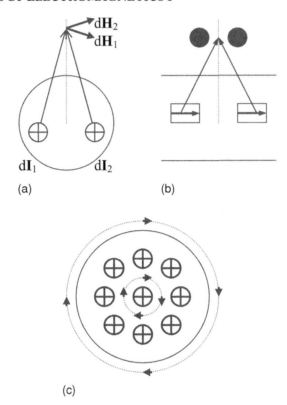

(a)

(b)

(c)

FIGURE 3.28: Infinitely long tube of uniform current: (a) angularly symmetric elements, (b) axially symmetric elements, and (c) Amperian paths.

located symmetrically with respect to the z-axis contribute equal angular components to the field, see Fig. 3.28(b). Moreover, the field has no variations in the z-direction since no matter at what point z an observation is made, the current stretches infinitely far on both sides. These considerations lead to the form of **H** as

$$\mathbf{H} = \mathbf{a}_\phi H(\rho). \tag{3.80}$$

Application of Ampere's law will yield the **H** field if we can select a suitable Amperian path on which $|\mathbf{H}| = 0$ or $|\mathbf{H}| = $ constant and on which **H** is either parallel or perpendicular to the line element dl. These are the same principles used to evaluate the fields from Gauss' law. An obvious path is a circular path of radius ρ centered on the z-axis as shown in Fig. 3.28(c). For this path, the LHS of Ampere's law is

$$\oint_L \mathbf{H} \cdot \mathbf{dl} = H_\phi 2\pi\rho. \tag{3.81}$$

The RHS of Ampere's law requires a bit of care. When $\rho \geq a$, all of the current within the tube is enclosed by the loop so that

$$I_{\text{ENC}} = \iint\limits_{S} \mathbf{J} \cdot \mathbf{ds} = \frac{I}{\pi a^2} \pi a^2 = I. \tag{3.82}$$

Equating these two results and solving for H_ϕ, we obtain

$$H_\phi = \frac{I}{2\pi\rho} \text{ A/m}, \quad \rho \geq a. \tag{3.83}$$

For $\rho \leq a$, the LHS of Ampere's law becomes

$$I_{\text{ENC}} = \iint\limits_{S} \mathbf{J} \cdot \mathbf{ds} = \frac{I}{\pi a^2} \pi \rho^2 = \frac{\rho^2}{a^2} I \tag{3.84}$$

The amount of current enclosed depends upon the radius of the path.
Setting Eqs. (3.81) and (3.84) equal and solving for H_ϕ, we obtain

$$H_\phi = \frac{\frac{\rho^2}{a^2} I}{2\pi\rho} = \frac{I\rho}{2\pi a^2} \text{ A/m}, \quad \rho \leq a. \tag{3.85}$$

The field intensity varies linearly with ρ inside the region of current and as ρ^{-1} outside of the current. A normalized plot of H_ϕ is shown in Fig. 3.29.

As shown above, application of Ampere's law works well for cylindrical geometries. It works well for planar current distributions, also. Consider the uniform surface current density $\mathbf{K} = K_0 \mathbf{a_y}$ A/m on the $z = 0$ plane shown in Fig. 3.30. From Fig. 3.30(a), it is clear that current elements located symmetrically with respect to the z-axis produce only an x-component of magnetic field. In addition, the fields above the current are oppositely directed to those

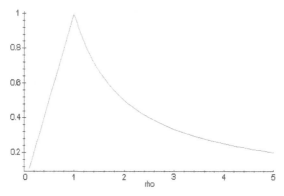

FIGURE 3.29: Magnetic field intensity of a uniform tube of current.

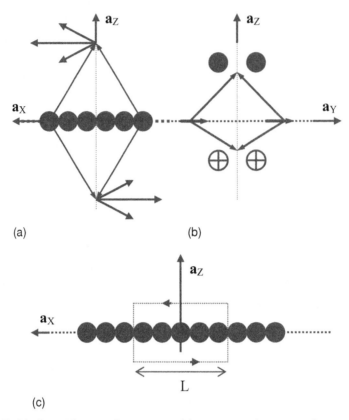

(a)   (b)

(c)

FIGURE 3.30: Field of a uniform surface current: (a) y-symmetric current elements, (b) x-symmetric current elements, and (c) Amperian path.

below in a manner consistent with the RH rule. Similar behavior for current elements located symmetrically with respect to the x-axis is pictured in Fig. 3.30(b). Furthermore, the doubly infinite extent of the current distribution guarantees that there are no variations of the field with x or y.

Since there is only an x-component of the magnetic field, a good Amperian path is a rectangular path as shown in Fig. 3.30(c); the magnetic field is aligned with the path for $z = $ constant portions of the path and is zero for the $x = $ constant portions of the path. With these facts, the RHS of Ampere's law gives

$$\oint_L \mathbf{H} \cdot \mathbf{dl} = H_X L + H_X L = 2 H_X L. \tag{3.86}$$

No matter the value of z, the only enclosed current is the surface current so that the RHS of Ampere's law is

$$I_{\text{ENC}} = \int\limits_{\text{BOUNDARY}} \mathbf{K} \cdot \mathbf{a}_N dl = \int\limits_{x=-L/2}^{L/2} K_O \mathbf{a}_Y \cdot \mathbf{a}_Y dx = K_O L. \qquad (3.87)$$

Equating Eqs. (3.86) and (3.87) and solving for H_X we obtain

$$H_X = \pm \frac{K_O}{2} \text{ A/m} \qquad (3.88)$$

where the $+$ sign applies for $z > 0$ and the $-$ sign for $z < 0$. The fields show no variation with respect to the distance from the current sheet. This is of the same form as electric fields due to a uniform planar surface charge.

   In summary, the magnetic fields of infinitely long, uniform current filaments are angularly directed around the current according to the RH rule and vary inversely with distance from the current axis. The magnetic fields of infinite, uniform current sheets are parallel to the current sheet, directed according to the RH rule, and have no variation with distance from the current sheet.

Example 3.15-1.   Calculate the fields of an infinite solenoid of radius $\rho = a$, that is located along the z-axis. The solenoid is composed of an infinitely long, uniform coil of N turns/meter. The current carries a current I. Due to the angular symmetry of the solenoid, the fields show no angular variations. Due to their infinite axial extent, they show no axial variation either. Figure 3.31(a) shows that the current establishes axially directed fields; those inside the solenoid are directed oppositely to those fields outside. Figure 3.31(b) shows two useful Amperian paths. Path 1 symmetrically encloses a section of the coil. Since the net current enclosed by Path 1 is zero and the fields outside are and in the same direction, the magnetic fields outside are zero. Path 2 is similar to Path 1 except that one axial segment of the path is within the solenoid. This is the only segment along which there is a nonzero magnetic field so the LHS of Ampere's law becomes $\oint_{L2} \mathbf{H} \cdot \mathbf{dl} = H_Z L$. The RHS of Ampere's law is $I_{\text{ENC}} = NIL$ where there are a total of NL turns within the path. Equating these results and solving for H_Z, we obtain $H_Z = NI$, not an unexpected result. This analysis reveals that the field is totally confined to the interior of the solenoid. In practical applications, there is leakage flux, but this is minimized by closely spacing the turns.

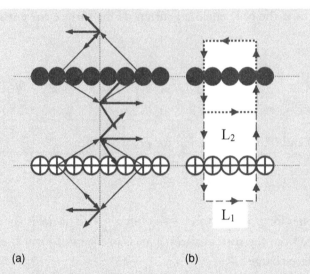

(a)   (b)

FIGURE 3.31: Infinite solenoid: (a) differential fields and (b) Amperian paths.

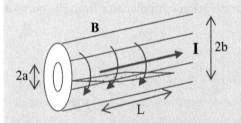

FIGURE 3.32: Coaxial cable.

Example 3.15-2. Calculate the inductance/meter of a coaxial cable. The cable has PEC inner and outer conductors of radii of a and b, respectively; see Fig. 3.32. A surface current is assumed to flow on the facing PEC surfaces. The current on the inner conductor is assumed to flow in the $\mathbf{a}_Z$ direction (to the right); the return current flows on the outer conductor in the $-\mathbf{a}_Z$ direction. The magnetic field within the cable is calculated by Ampere's law as $H_\phi = I/2\pi\rho$. If we consider only 1 m length of cable with an ideal short circuit on one end, then a "loop" of current is formed by the inner conductor, the short circuit, and the outer conductor. This configuration certainly differs from a planar wire loop, but it forms a closed circuit that encloses a magnetic flux. The flux enclosed by the 1-m cable is given by

$$\Psi_m = \iint_S \mathbf{B} \cdot \mathbf{ds} = \int_{z=0}^{1} \int_{\rho=a}^{b} \mu H_\phi \mathbf{a}_\phi \cdot \mathbf{a}_\phi \, d\rho \, dz$$

$$= \int_{z=0}^{1} \int_{\rho=a}^{b} \frac{\mu I}{2\pi\rho} \, d\rho \, dz = \frac{\mu I}{2\pi} \ln\left(\frac{b}{a}\right).$$

Since this is the flux in 1 m length of cable the inductance/meter of the coaxial cable is given as

$$L/m = \frac{\Psi_m}{I} = \frac{\mu}{2\pi} \ln\left(\frac{b}{a}\right) \text{ H/m.}$$

Example 3.15-3. Calculate the internal inductance/meter of a wire of radius a with uniform current density. The *internal inductance* is defined as that portion of inductance due to fields that exist within the conductor. The magnetic field within a long wire with a uniform current density is given by Eq. (3.85) as $H_\phi = \frac{(\rho^2/a^2)I}{2\pi\rho} = \frac{I\rho}{2\pi a^2}$. Since the fields of interest are known and confined to a limited region, calculation of inductance is easiest using energy concepts as in Eq. (3.50),

$$L = \frac{2W_m}{I_L^2} = \frac{2 \iiint\limits_{\forall} \frac{\mu|\mathbf{H}|^2}{2} dv}{\left(\int\limits_L \mathbf{H} \cdot \mathbf{dl}\right)^2}$$

$$= \frac{\int\limits_{z=0}^{1} \int\limits_{\phi=0}^{2\pi} \int\limits_{\rho=0}^{a} \mu \frac{I^2\rho^2}{4\pi^2 a^4} \rho d\rho d\phi dz}{\left(\int\limits_{\phi=0}^{2\pi} \frac{I\rho}{2\pi a^2} \rho d\phi \Big|_{\rho=a}\right)^2} = \frac{\mu}{8\pi} \text{ H/m.}$$

No matter what the diameter of the wire, the internal inductance is the same. Most conductive wires have $\mu = \mu_o$ and the internal inductance is 5 μH/cm.

Example 3.15-4. Calculate the mutual inductance between an infinite filament of uniform current and an N-turn rectangular coil of width w and length L located at a distance d from the filament, see Fig. 3.33. The field of the filament is given by $H_\phi = I_1/2\pi\rho$ so that the flux that penetrates the rectangular coil is given as

$$\Psi_{m2} = \iint\limits_{S_2} \mathbf{B}_1 \cdot \mathbf{ds} = \int\limits_{z=0}^{L} \int\limits_{\rho=d}^{d+w} \frac{\mu I_1}{2\pi\rho} d\rho dz = \frac{\mu I_1 L}{2\pi} \ln\left(\frac{d+w}{d}\right).$$

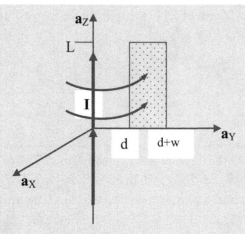

FIGURE 3.33: Mutual inductance between an infinitely long filament and a nearby coil.

The mutual inductance is then calculated as

$$M = \frac{N_2 \Psi_{m2}}{I_1} = \frac{\mu N_2 L}{2\pi} \ln\left(\frac{d+w}{d}\right) \ H.$$

More sophisticated inductance concepts and calculations will not be covered here. However, other magnetic devices are studied in the next chapter.

CHAPTER 4

Magnetic Devices

Energy conversion and transmission processes employ a variety of magnetic devices. Several new ideas will be added to the principles covered in the previous chapter to describe their operation. Specialization of these concepts to magnetic devices is the focus of this chapter. In addition, a more general interpretation of Faraday's law concludes this chapter.

4.1   MAGNETIC CIRCUITS: THE BASICS

The combination of a magnetic core and an excitation coil is known as a *magnetic circuit*. A wide variety of magnetic devices can be described by different configurations of this basic magnetic circuit. This section focuses upon calculation of the fields within a magnetic circuit and upon their behavior.

Reconsider the toroidal, linear, magnetic core of rectangular cross section with permeability μ excited by an N-turn coil carrying a current I as introduced in Section 3.4 and shown again in Fig. 4.1.

From Ampere's law, the magnetic field intensity is calculated from as

$$H_\phi = \frac{NI}{2\pi\rho}. \tag{4.1}$$

Within the core, the flux density is given by

$$B_\phi = \mu H_\phi = \frac{\mu NI}{2\pi\rho} \tag{4.2}$$

and the total flux within the core is determined by integrating B_ϕ over the cross section of the core as

$$\Psi_m = \iint\limits_A \mathbf{B} \cdot \mathbf{ds} = \int\limits_{z=z_1}^{z_1+w} \int\limits_{\rho=a}^{b} \frac{\mu NI}{2\pi\rho}\mathbf{a}_\phi \cdot \mathbf{a}_\phi d\rho dz$$

$$= \frac{\mu NI}{2\pi} w \ln\left(\frac{b}{a}\right). \tag{4.3}$$

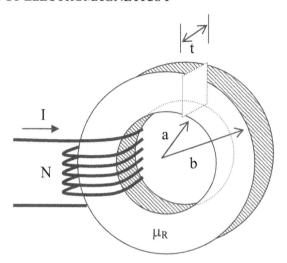

FIGURE 4.1: Magnetic circuit of a toroidal core.

These computations are relatively easy for this simple configuration because it aligns exactly with the cylindrical coordinate system. However, more complex structures do not fit so nicely and become computational nightmares. The usual strategy for reasonably accurate results without laborious calculations is to make simplifying, yet fairly accurate approximations. From Eq. (3.11), **H** and **B** are seen to be greatest at the inner radius of the core and least at the outer. *The simplest approximation is that the fields are uniform throughout every cross section of the core and to use the average or mean value of the path length for their calculation.* Fortunately, this approximation is especially accurate for the fields and total flux when the radial thickness of the core is significantly less than its mean circumference. In fact, this approximation is within 10% of the exact value for the radial thickness equal to the mean radius. It works so well that we will use this for nearly all configurations of cores with any cross section. Application of this approximation to the magnetic field intensity calculations gives

$$\oint_L \mathbf{H} \cdot \mathbf{dl} \approx H_\phi l_{\text{MEAN}} = NI = I_{\text{ENC}}. \qquad (4.4)$$

Since we know that the flux is directed angularly within the core, the subscript on the field is not needed. As a general convention, the mean path length is always used, so we will drop that subscript as well. Solving for the fields we have

$$H = \frac{NI}{l} = \frac{NI}{2\pi \rho_{\text{MEAN}}} = \frac{NI}{\pi (a + b)} \qquad (4.5)$$

and

$$B = \frac{\mu NI}{\pi(a+b)}. \tag{4.6}$$

The mean path length approximation is equivalent to requiring uniform H and B throughout each cross section. Since B is uniform, the flux is found quite simply as

$$\Psi_m = \iint_A \mathbf{B} \cdot \mathbf{ds} = BA = \frac{\mu NI(b-a)w}{\pi(a+b)}. \tag{4.7}$$

Example 4.1-1. Calculate the exact fields and the total flux in the toroidal core of Fig. 4.1 for $a = 3$ cm, $b = 5$ cm, and $t = 2$ cm, with $\mu_R = 1000$, and with a coil of 100 turns carrying 1 A of current. Exact calculations require use of Eqs. (3.11), (3.12), and (3.25) which give

$$H_\phi = \frac{NI}{2\pi\rho} = \frac{100(1)}{2\pi\rho} = \frac{50}{\pi\rho} \text{ A/m},$$

$$B_\phi = \frac{\mu NI}{2\pi\rho} = \frac{4\pi 10^{-7}(10^3)(100)(1)}{2\pi\rho} = \frac{0.02}{\rho} \text{ wb/m}^2,$$

and

$$\Psi_m = \iint_A \mathbf{B} \cdot \mathbf{ds} = (0.02)\ln\left(\frac{0.05}{0.03}\right)(0.02) = 204 \ \mu\text{wb}.$$

Example 4.1-2. Calculate the fields and flux of Example 4.1-1 using the approximate expressions of Eqs. (4.1), (4.2), and (4.4) $H = \frac{NI}{l} = \frac{NI}{\pi(a+b)} = \frac{100}{0.08\pi} = \frac{1250}{\pi}$ A/m, $B = \mu H = 4\pi 10^{-4}\frac{1250}{\pi} = 0.5$ wb/m^2, and $\Psi_m = BA = (0.5)(0.02)(0.02) = 200 \ \mu$wb. The values of B and H are the same as the exact values (evaluated at the mean radius, $\rho = 0.04$ m) of Example 4.1-1. Though Ψ_m is an approximation assuming uniform flux density, it is within 2% of the exact value of Example 4.1-1, not bad.

The magnetic core is often composed of more than one magnetic material. The continuity of magnetic flux in the core of Fig. 4.2(a) requires that all of the flux leaving one material enter into the other material at the boundary between the two materials. Since the flux densities are uniform and the areas the same, the flux densities are equal as well so that

$$B_1 = B_2 \quad \text{or} \quad \mu_1 H_1 = \mu_2 H_2. \tag{4.8}$$

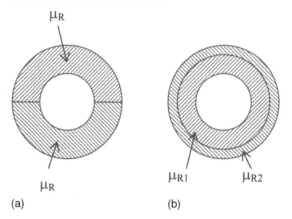

FIGURE 4.2: Composite magnetic circuits: (a) series flux and (b) parallel flux.

The difference in permeabilities requires that H_1 and H_2 are different. Application of Ampere's law gives

$$\oint_L \mathbf{H} \cdot \mathbf{dl} = H_1 l_1 + H_2 l_2 = NI. \tag{4.9}$$

Combining Eqs. (4.10) and (4.11), we solve for H_1 and H_2 as

$$H_1 = \frac{NI}{l_1 + l_2 \frac{\mu_2}{\mu_1}} \quad \text{and} \quad H_2 = \frac{NI}{l_2 + l_1 \frac{\mu_1}{\mu_2}} \tag{4.12}$$

and B_1 and B_2 as

$$B_1 = \frac{\mu_1 NI}{l_1 + l_2 \frac{\mu_1}{\mu_2}} = \frac{NI}{\frac{l_1}{\mu_1} + \frac{l_2}{\mu_2}} \quad \text{and} \quad B_2 = \frac{\mu_2 NI}{l_2 + l_1 \frac{\mu_2}{\mu_1}} = \frac{NI}{\frac{l_1}{\mu_1} + \frac{l_2}{\mu_2}}. \tag{4.13}$$

The fluxes in both regions are the same and are given by

$$\Psi_m = B_1 A = B_2 A = \frac{NIA}{\frac{l_2}{\mu_2} + \frac{l_1}{\mu_1}}. \tag{4.14}$$

This procedure is valid for k different core materials in series and is governed by

$$H_1 l_1 + H_2 l_2 + \cdots + H_k l_k = \sum_{i=1}^{k} H_i l_i = NI$$
$$\Psi_m = B_1 A = B_2 A = \cdots = B_k A$$
$$B_1 = \mu_1 H_1, \, B_2 = \mu_2 H_2, \ldots, \, B_k = \mu_k H_k. \tag{4.15}$$

Another form of composite magnetic circuits is shown in Fig. 4.2(b) where the flux has two or more alternate paths. This problem can be approached as two separate problems with the same source of magnetic field, NI, for which expressions for the magnetic field can be written as

$$H_1 l_1 = NI = H_2 l_2. \tag{4.16}$$

The flux paths are often different lengths, $l_1 \neq l_2$, which results in $H_1 \neq H_2$ for the uniform flux assumption. Note that this violates the boundary conditions of equal tangential components of the magnetic field intensity at the material interface. Alternatively, we can match the boundary conditions, $H_1 = H_2$, which requires $l_1 = l_2$, a condition which may not match the core geometry. This problem arises from the approximation of equal path length for all flux within a branch of the core. Either condition may be used; both represent a good approximation. The continuity of flux leads to

$$\Psi_{m1} = B_1 A_1 \quad \text{and} \quad \Psi_{m2} = B_2 A_2 \tag{4.17}$$

and of course the constitutive relations

$$B_1 = \mu_1 H_1 \quad \text{and} \quad B_2 = \mu_2 H_2. \tag{4.18}$$

H is readily calculated as

$$H = H_1 = H_2 = \frac{NI}{l} \tag{4.19}$$

from which the B_1 and B_2 follow as

$$B_1 = \frac{\mu_1 NI}{l} \quad \text{and} \quad B_2 = \frac{\mu_2 NI}{l}. \tag{4.20}$$

The total flux in the core is given as

$$\Psi_m = \Psi_{m1} + \Psi_{m2} = B_1 A_1 + B_2 A_2$$
$$= NI \left(\frac{\mu_1 A_1 + \mu_2 A_2}{l} \right). \tag{4.21}$$

A generalization to k different materials in a configuration as in Fig. 4.2(b) gives

$$H_1 = H_2 = \cdots = H_k = H = \frac{NI}{l}$$

$$\Psi_m = B_1 A_1 + B_2 A_2 + \cdots + B_M A_M = \sum_{i=1}^{k} B_i A_i$$

$$B_1 = \mu_1 H, \ B_2 = \mu_2 H, \ \ldots, \ B_k = \mu_k H. \tag{4.22}$$

The concept of series or parallel flux paths can be extended readily to configurations where the cross-sectional areas or flux path lengths differ with minor algebraic modifications. The governing principles are that the flux is constant through a series connection and that the mmf drop is the same across a parallel connection.

Example 4.1-3.   Calculate the flux present in the core of Example 4.1-1 when one semi-circular half of the core has $\mu_R = 1000$ and the other half has $\mu_R = 2000$. Since the flux through both halves is the same, $\Psi_{m1} = \Psi_{m2}$, the equality of the areas requires that $B_1 = B_2$ which leads to $\mu_1 H_1 = \mu_2 H_2$. The total mmf drop must equal the sum of mmf drops of the two halves or $NI = (H_1 + H_2)l$ since the two halves have equal length. The simultaneous solution of these two equations gives

$$H_1 = \frac{NI}{l\left(1 + \frac{\mu_2}{\mu_1}\right)} \quad \text{and} \quad H_2 = \frac{NI}{l\left(1 + \frac{\mu_1}{\mu_2}\right)}.$$

The total flux is

$$\Psi_m = B_1 A = B_2 A = \frac{NIA}{\ell\left(\frac{1}{\mu_2} + \frac{1}{\mu_1}\right)} = \frac{100(0.02)(0.02)}{\frac{0.08\pi}{4\pi\times10^{-7}}\left(\frac{1}{1000} + \frac{1}{2000}\right)} = 133 \ \mu\text{wb}.$$

Example 4.1-4.   Calculate the flux present in the core of Example 4.1-1 when the core is divided radially into two equally thick, concentric regions, the inner region, $0.03 \leq \rho \leq 0.04$, has $\mu_R = 1000$, the outer region $0.04 \leq \rho \leq 0.05$, has $\mu_R = 2000$. Since the coil encircles both regions, they both have $NI = 100$ At of mmf drop with resulting magnetic field intensities of

$$H_{\text{INNER}} = \frac{NI}{l_{\text{INNER}}} = \frac{100}{0.035(2\pi)} = 454.8$$

$$H_{\text{OUTER}} = \frac{NI}{l_{\text{OUTER}}} = \frac{100}{0.045(2\pi)} = 353.7.$$

Note that the path in each of the region is approximated as the mean length in that region. The total flux in the core is the sum of the fluxes in the two regions as

$$\begin{aligned}
\Psi_m &= \Psi_{m1} + \Psi_{m2} = \mu_1 H_1 A_1 + \mu_2 H_2 A_2 \\
&= 4\pi(10^{-7})[10^3(454.8) + 2(10^3)(353.7)](0.0002) \\
&= 292 \ \mu\text{wb}.
\end{aligned}$$

Alternatively, let's assume that $l_1 = l_2 = 2\pi(0.04) = 0.251$ m which leads to $H_1 = H_2 = NI/l = 100/(0.251) = 398.4$ At. Note that this value of H is approximately the average of the two values obtained by the first method, $H_{AVE} = (H_1 + H_2)/2 = 404.25$. The flux is $\Psi_m = 4\pi(10^{-7})[10^3 + 2(10^3)](398.4)(0.0002) = 300$ μwb. The two results differ by only 3%.

4.2 MAGNETIC CIRCUITS: MORE DETAILS

We covered the basic principles of magnetic circuits in the last section. Now we can look a little deeper at some special cases and make some interpretations of the results.

The flux established in a magnetic circuit is proportional to the source NI. The flux within a branch is continuous and depends upon the geometry of the branch. The sum of Hl around a path is equal to the source NI. This seems very much like the language of circuits where the current is analogous to the flux, the voltage source is analogous to NI, and the geometry of a branch relates the voltage and current analogous to the relationship of NI and the flux. Let's look at this idea more carefully. Consider the magnetic circuit and analogous electric circuit of Figs. 4.3(a) and (b).

The KVL from electric circuits states that for a single source in a closed loop, the sum of the resistive voltage drops is equal to the voltage source,

$$V_S = \sum_{i=1}^{N} V_i. \tag{4.23}$$

Ampere's law for a closed magnetic circuit takes a similar form

$$NI = \sum_{i=1}^{N} H_i l_i = \sum_{i=1}^{N} V_{mi}. \tag{4.24}$$

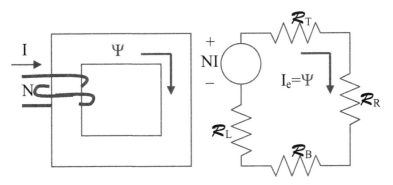

FIGURE 4.3: Analogous circuits: (a) magnetic circuit and (b) electric circuit.

The similarity of these forms suggests that the magnetic source, NI, corresponds to the electric voltage source, V_S. The individual resistive voltage drops, V_i, correspond to mmf drops, $H_i l_i = V_{mi}$. The magnetic circuit equivalent of KVL is that the magnetic source establishes an mmf potential, NI, across the magnetic circuit with each branch in the loop possessing an mmf drop or magnetic potential drop, $V_{mi} = H_i l_i$. The longer the branch lengths, l_i, the larger the source, NI, needed to establish a given H.

The KCL of electric circuits states that in a closed loop the currents that flow into and out of every branch are the same,

$$I_1 = I_2 = \cdots = I_M = I. \qquad (4.25)$$

The flux entering and leaving each branch of a closed magnetic circuit are equal,

$$\Psi_{m1} = \Psi_{m2} = \cdots = \Psi_{mM} = \Psi_m. \qquad (4.26)$$

Obviously, the magnetic flux is analogous to the current flux in resistive circuits. When there are N branches at a circuit node, the sum of the currents entering or leaving must be zero,

$$\sum_{i=1}^{N} I_i = 0. \qquad (4.27)$$

Analogously, when there are N branches at a junction of a magnetic circuit, the sum of the fluxes entering or leaving must be zero,

$$\sum_{i=1}^{N} \Psi_{mi} = 0. \qquad (4.28)$$

Extending the analogy to its logical conclusion, we first note that the ratio of a branch voltage drop, V_i, to a branch current, I_i, gives the branch resistance, R_i. Analogously, the ratio of a branch mmf drop, V_{mi}, to the branch flux, Ψ_{mi}, is a sort of magnetic resistance. This is the *reluctance* that we encountered in Chapter 3; it is denoted by $\mathcal{R}$ with units of Henries^{-1}. The reluctance of the ith branch is given by

$$\mathcal{R}_i = \frac{V_{mi}}{\Psi_{mi}} = \frac{H_i l_i}{\Psi_{mi}}. \qquad (4.29)$$

The geometrical dependence of reluctance can be obtained from the parameters of a toroidal core, $V_m = Hl = NI$ and $\Psi_m = \mu NIA/l$, as

$$\mathcal{R}_i = \frac{H_i l_i}{\Psi_{mi}} = \frac{NI}{\frac{\mu_i NIA_i}{l_i}} = \frac{l_i}{\mu_i A_i}. \qquad (4.30)$$

This is amazing! The analogy of reluctance with resistance holds even in the form of geometrical dependence (you do recall $R = L/\sigma A$, don't you?). The longer the element, the greater the resistance (reluctance), and the greater the material's conductivity (permeability) or its cross-sectional area, the smaller the resistance (reluctance). A small reluctance is desired in order to establish a large flux with as small a NI as possible. From Eq. (4.30), we see that a short path, a large cross section, and, most importantly, a high permeability are necessary to achieve small reluctance. Fortunately, this expression is quite general and can be applied to most geometries when we assume uniform flux density and using mean flux path.

This analog enables us to use our experience with electric circuits to understand magnetic circuits. With only simple reluctance calculations via Eq. (4.30), we can see quickly which branches will present a high mmf or which branches enable a large flux to exist. In addition, it enables us to perform series and parallel calculations for complicated magnetic circuits. For example, a composite circuit with two parallel flux paths can be represented by a parallel circuit with reluctances of

$$\mathcal{R}_1 = \frac{l_1}{\mu_1 A_1} \quad \text{and} \quad \mathcal{R}_2 = \frac{l_2}{\mu_2 A_2}. \tag{4.31}$$

The combined reluctance for the parallel flux paths is the sum of the reciprocals, or

$$\mathcal{R}_{\text{TOT}} = \left(\mathcal{R}_1^{-1} + \mathcal{R}_2^{-1}\right)^{-1} = \left(\frac{\mu_1 A_1}{l_1} + \frac{\mu_2 A_2}{l_2}\right)^{-1}. \tag{4.32}$$

The total flux in the core is simply calculated as

$$\Psi_m = NI\mathcal{R}_{\text{TOT}}^{-1} = NI\left(\frac{\mu_1 A_1}{l_1} + \frac{\mu_2 A_2}{l_2}\right), \tag{4.33}$$

which is identical to Eq. (4.14) derived directly from field concepts.

There is a minor problem in implementing this concept for a junction with more than two branches. Consider three branches that join together as shown in Fig. 4.4.

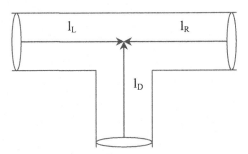

FIGURE 4.4: Magnetic circuit junction geometry.

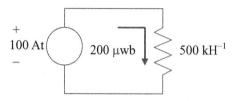

FIGURE 4.5: Electric circuit analog.

The flux paths in the vicinity of the junction are quite complex, depending upon the cross-sectional areas of all three branches. However, a *simplifying approximation is to define the lengths of the branches from a common point at the center of the junction.* Experience has shown that reasonable accuracy is obtained with no further assumptions.

Example 4.2-1. Sketch the circuit analog of the magnetic circuit of Example 4.1-1. The mmf source has a value of $NI = 100$. The reluctance of the core is calculated as

$$R = \frac{l}{\mu A} = \frac{0.04(2\pi)}{4\pi \times 10^{-4}(0.0004)} = 500{,}000 \text{ H}^{-1}.$$

The flux is $\Psi_m = NI/R = 100/500{,}000 = 200 \text{ μwb}$.

Example 4.2-2. Calculate the reluctance of the core of Example 4.1-3. The total reluctance is the sum of the reluctances of the two halves as

$$R = R_2 + R_1 = \frac{l_1}{\mu_1 A_1} + \frac{l_2}{\mu_2 A_2}$$

$$= \frac{0.04\pi}{4\pi(10^{-4})(0.0004)} + \frac{0.04\pi}{4\pi(2)(10^{-4})(0.0004)}$$

$$= 375{,}000 \text{ H}^{-1}.$$

Example 4.2-3. Calculate the reluctance of the core of Example 4.1-4. The total reluctance is the parallel reluctances of the two separate regions as

$$R_{\text{TOT}} = \left(R_1^{-1} + R_2^{-1}\right)^{-1} = \left(\frac{\mu_1 A_1}{l_1} + \frac{\mu_2 A_2}{l_2}\right)^{-1}$$

$$= 4\pi \times 10^{-7}\left(\frac{10^3}{0.035(2\pi)} + \frac{2 \times 10^3}{0.045(2\pi)}\right)(0.0002)$$

$$= 342{,}391 \text{ H}^{-1}.$$

4.3 MAGNETIC CIRCUITS WITH AIR GAPS

A common magnetic circuit is that of a motor which has a small air gap between the fixed stator and the moveable rotor. The magnetic flux is established by current flow in coils that are wound on the magnetic material of the stator and/or rotor. The flux bridges the gap between the adjoining iron regions known as pole pieces. But, from Eq. (4.30), we see that the very small value of the permeability of the air gap, μ_o, results in a very large air gap reluctance which in turn requires a large NI to establish the necessary flux. This is strong motivation for keeping the air gap as small as possible without the rotor and stator touching. Even with a very small gap, its reluctance is much larger than the reluctance of the rotor and stator combined. This leads to another important approximation in magnetic circuits: *as a first approximation for a branch that contains a series air gap, the total reluctance of the branch is that of the air gap.* Of course, more accurate calculations must include the reluctance of the remainder of the branch.

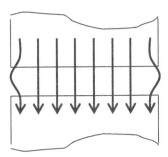

FIGURE 4.6:   Fringing at an air gap.

Example 4.3-1.   Consider Example 4.1-1 with a modified core to include a 1 mm radial air gap. The reluctance of the core and the gap are

$$\mathcal{R}_C = \frac{l_C}{\mu_C A_C} = \frac{2\pi(0.04)}{4\pi 10^{-7}(10^3)(0.02)(0.02)} = 0.5 \times 10^6 \text{ H}^{-1}$$

and

$$\mathcal{R}_G = \frac{l_G}{\mu_G A_G} = \frac{10^{-3}}{4\pi 10^{-7}(0.02)(0.02)} = 1.99 \times 10^6 \text{ H}^{-1},$$

respectively, with a total reluctance of $\mathcal{R}_T = \mathcal{R}_C + \mathcal{R}_G = 2.49 \times 10^6 \text{H}^{-1}$. The flux is calculated easily as $\Psi_m = \frac{Hl}{\mathcal{R}_i} = \frac{NI}{\mathcal{R}} = \frac{100}{2.49 \times 10^6} = 40.2$ μwb compared to 204 μwb without the gap. The flux is greatly reduced due to the greatly increased reluctance of the circuit due to the air gap. When the reluctance of the circuit is approximated by the reluctance of the gap only, the flux is 50.3 μwb, a 25% error. In most cases the error will be less.

The flux through the air gap is continuous and equal to the flux in the core. But in the absence of high-μ material in the gap, the flux is not as well guided as in the core. This allows the flux to "bulge" slightly out of the gap, *fringing* around the outer edges of the pole pieces, see Fig. 4.6. A common rule of thumb is that the "effective" area of the gap is 1.1 times the area of the core to account for the fringing. Of course, the flux density in the gap is reduced by 1.1. This approximation is fairly accurate as long as the gap length is much less than the smallest dimension of the adjoining pole pieces. Since our goal is to understand principles and not to strive for the ultimate accuracy, we will ignore this slight increase in the area of the flux path through an air gap.

4.4   NONLINEAR MAGNETIC CIRCUITS

Up to this point, we have assumed that the magnetic material is linear. However, for large magnetic fields often required in power applications, the permeability of most materials shows a nonlinear dependence on the magnetic field intensity, i.e., $\mu = \mu(H)$. This nonlinear behavior is often displayed graphically as a plot of B versus H, see Fig. 4.7. For linear material $B = \mu H$ is represented by a straight line through the origin with a constant slope of μ. Nonlinear magnetic materials are nearly linear for low level fields, but are characterized by a saturation of B as H becomes large. The causes of this behavior are described briefly in Section 3.5. Ampere's law, flux continuity, and $B = \mu H$ govern all magnetic behavior whether linear or nonlinear materials are involved. But the solution process is complicated by the nonlinear nature of the permeability. When there is only a single flux path in the core, techniques used with linear cores can be readily adapted. When the excitation current, I, is given, Ampere's law is used to determine H that in turn is used to determine B from the B–H curve from which the flux is calculated. When the required flux is given, B is calculated by flux continuity which in turn provides the required H from the B–H curve and the excitation current follows from Ampere's law. However, these solution methods become much more complicated or fail when there are multiple flux paths.

Several strategies which work well for "ball park" answers to single flux path problems are considered in the following paragraphs. We will defer the more advanced methods required for multiple flux paths until a more advanced course.

Table lookup or curve-fitting methods make use of tables of values or mathematical functions via curve-fitting techniques to approximate the B–H curve behavior. Numeric solutions of the resulting equations are well suited to software packages such as MAPLE and MATHEMATICA. The solutions depend strongly on the accuracy of the curve fit, a typical B–H curve as shown in Fig. 4.7 can require a 15th order polynomial for good approximation.

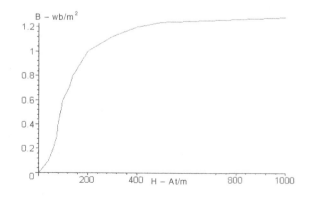

FIGURE 4.7:   B–H curve for nonlinear material.

Circuit analogs using PSPICE are useful once the B–H curve polynomial has been determined. The dimensions of magnetic circuit elements and the polynomial specification for controlled sources can be used to simulate the nonlinear core behavior. In addition, the PSPICE *K* device type provides direct specification of area, core length, and gap length of inductors. The core material is selected from a library of commercial materials or designed by the user.

A *graphical method*, based upon the analog between electric and magnetic circuits, provides reasonable accuracy rather quickly without the necessity of a computer. The B–H behavior of the linear portion of the circuit is a straight line that can be determined from its analogous Thévenin equivalent circuit. The solution is the point at which this linear curve and the B–H curve intersect.

A typical engineering strategy—make a *piecewise linear (PWL) approximation* to the B–H curve—is the basis for another solution method. The B–H curve is approximated by suitable straight-line segments; the accuracy of the solution dictates the number of PWL segments. The solution is assumed to lie within a likely PWL region and linear analysis is applied. If the solution does not exist in this region, the process is repeated for another region suggested by the original solution results until a consistent solution is obtained.

A *"cut and try"* method assumes a solution point and solves for the mmf drops via the $B = \mu H$ relationship directly from the B–H curve. This is compared with the mmf drops calculated directly from the assumed solution value of H. If the two results do not compare closely enough, then an increased or decreased H as suggested by the results is the basis for a second round of calculations. This process is continued until sufficient accuracy is obtained.

Example 4.4-1. The core of Example 4.1-1 has material with the properties exhibited in Fig. 4.7. Calculate the flux by table lookup method. Ampere's law gives $H = NI/l = 100/0.08\pi = 397.9$ At/m. From Fig. 4.7, this corresponds to $B = 1.2$ wb/m^2. The flux in the core is given as $\Psi_m = BA = 1.2(0.0002) = 240$ µwb.

Example 4.4-2. The core of Example 4.4-1 has a 0.5 mm radial air gap added. Calculate the current needed to produce an air gap flux density of 1 wb/m^2 by table lookup. The mmf drop of the air gap is calculated as $H_G l_G = (1/4\pi \times 10^{-7})(0.0005) = 397.9$ At. From the B–H curve of Fig. 4.7, the field intensity in the core is given by $H_C = 400$ At/m and the mmf drop in the core is given by $V_{mC} = 400(0.08\pi) = 100.5$ At. The total mmf drop for the circuit is $NI = \sum H_i l_i = 489.4$ At. The required current is $I = 4.894$ A. Note that the mmf drop of the air gap is nearly four times that of the core.

Example 4.4-3. The core with a gap of Example 4.4-2 has an excitation current of 3 A. Calculate the air gap flux density by the graphical method. The nonlinear core behaves according to the B–H curve of Fig. 4.7. Utilizing $B_C = B_G$ and substituting for H_G from Ampere's law, we obtain

$$B_C = B_G = \mu_0 H_G = \frac{\mu_0}{l_G}(NI - H_C l_C) = -H_C l_C \frac{\mu_0}{l_G} + NI \frac{\mu_0}{l_G}$$

$$= -H_C \left(\frac{(0.08\pi)4\pi \times 10^{-7}}{0.0005}\right) + \frac{4\pi \times 10^{-7}}{0.0005} 300$$

$$= -(6.32 \times 10^{-4})H_C + 0.754,$$

which is of the form $y = mx + b$ in terms of B_C and H_C. The x-intercept is $0.754/6.32 \times 10^{-4} = 1193$; the y-intercept is $b = 0.754$. This linear form represents the "terminal" behavior of the NI source and the $H_G l_G$ of the air gap. A plot of this straight line will intersect the B–H curve at the point where $B_C = B_G$ and provides the operating point of the magnetic circuit, see Fig. 4.8. From Fig. 4.8, the solution is $B_G = 0.68$ wb/m^2.

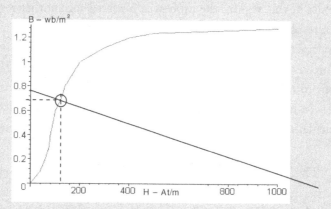

FIGURE 4.8: Graphical nonlinear solution method.

Example 4.4-4. Solve Example 4.4-3 by the cut and try method. Try a value of $B_C = B_G = 1$ wb/m^2 which gives $H_C = 200$ A/m and $H_G = 795, 775$ A/m and $H_C l_C = 50.3$ At and $H_G l_G = 397.9$ At. This requires $NI = 448 > 300$ At. So we try a smaller value of $B_C = B_G = 0.8$ wb/m^2 accompanied by $H_C = 130$ At and $H_G = 636, 620$ and $H_C l_C = 32.7$ and $H_G l_G = 318.3$ with $NI = 351 > 300$ At. Another try of $B_C = B_G = 0.7$ wb/m^2 gives $H_C = 120$ At and $H_G = 557, 042$ At with $H_C l_C = 30.2$ At and $H_G l_G = 278.5$ At. The total mmf drop is $NI = 308.7 \approx 300$ At, close enough. So we obtain $B_C \approx 0.7$ wb/m^2, very close to the graphical results of Example 4.4-3.

4.5   TRANSFORMERS

The induction of a voltage in one coil by a time-varying current in second coil offers the possibility of transferring power from one portion of a circuit to another without a direct connection. The flow of power is through the flux generated by one coil which links one or more other coils. Transformers are magnetic devices that are especially designed for this function. Transformers take on many different shapes and sizes depending upon their application. However, we will consider only a single-branched, closed magnetic core with two coils from which we can obtain the basic principles.

When two or more coils are wound on a ferromagnetic core with a large relative permeability, there is insignificant fringing of the magnetic fields outside of the core. Both coils are linked by the same flux and the coupling coefficient is unity, i.e., $k = 1$. In many applications, one of the coils is designed for application of the excitation signal; this coil functions as the transformer input and is called the *primary coil, primary winding,* or just *primary.* The other coil which provides the output is known as the *secondary coil, secondary winding,* or *secondary.* We will assume that the coils are made of PEC wire so that both coils have zero resistance. In actual transformers, the diameter of the wire is large enough to keep the resistance acceptably low, but small to fit through the center of the core. See Fig. 4.9 for a diagram of an elementary transformer.

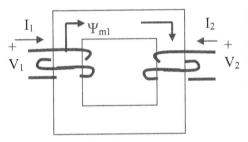

FIGURE 4.9:   Elementary transformer.

The voltage drop across the primary is due to two separate effects—the self-induced voltage due to the flux excited by the current in the primary and a mutually coupled voltage due to the flux excited by the current in the secondary. Since the core is linear, the total voltage is the sum of these two voltages as

$$V_1 = \frac{d N_1 \Psi_1}{dt} + \frac{d N_1 \Psi_2}{dt} = N_1 \frac{d\Psi_1}{dt} + N_1 \frac{d\Psi_2}{dt}. \qquad (4.34)$$

Using Eqs. (3.33) and (3.55), we note that

$$\frac{d N_1 \Psi_1}{dt} = \frac{d\left(\frac{N_1 \Psi_1}{I_1} I_1\right)}{dt} = \frac{d(L_1 I_1)}{dt} = L_1 \frac{d I_1}{dt} \qquad (4.35)$$

and

$$\frac{d N_1 \Psi_2}{dt} = \frac{d\left(\frac{N_1 \Psi_2}{I_2} I_2\right)}{dt} = \frac{d(M I_2)}{dt} = M \frac{d I_2}{dt}. \qquad (4.36)$$

This leads to the circuit form for the voltage across the primary as

$$V_1 = L_1 \frac{d I_1}{dt} + M \frac{d I_2}{dt}.$$

(4.37)

Similarly, the voltage drop across the secondary is expressed in terms of the fluxes as

$$V_2 = \frac{d N_2 \Psi_1}{dt} + \frac{d N_2 \Psi_2}{dt} = N_2 \frac{d \Psi_1}{dt} + N_2 \frac{d \Psi_2}{dt}$$

(4.38)

and in circuit form as

$$V_2 = L_2 \frac{d I_2}{dt} + M \frac{d I_1}{dt}.$$

(4.39)

FIGURE 4.10: Circuit model for transformer.

L_1 and L_2 are the self-inductance of the primary and the secondary, respectively, and M is the mutual inductance between the primary and the secondary coils. The sign of M is positive for the core in Fig. 4.9 since the fluxes are similarly directed for the positively defined currents. If the fluxes are oppositely directed, the sign of M is negative; more on this is given later. Equations (4.37) and (4.39) are the basis for the two-port circuit model for a transformer as shown in Fig. 4.10. A note of caution, this model is valid for time-varying currents only; DC currents cannot pass through a transformer.

Since there is no fringing, the flux is the same everywhere throughout the core and is equal to the sum of the two fluxes, i.e., $\Psi_{CORE} = \Psi_1 + \Psi_2$. With this relationship, Eqs. (4.34) and (4.38) are rewritten in the form

$$\begin{aligned} V_1 &= \frac{d N_1 \Psi_1}{dt} + \frac{d N_1 \Psi_2}{dt} = N_1 \frac{d (\Psi_1 + \Psi_2)}{dt} \\ &= N_1 \frac{d \Psi_{CORE}}{dt} \end{aligned}$$

(4.40)

and

$$V_2 = \frac{d N_2 \Psi_1}{dt} + \frac{d N_2 \Psi_2}{dt} = N_2 \frac{d \Psi_{CORE}}{dt},$$

(4.41)

which leads to

$$\frac{V_1}{V_2} = \frac{N_1}{N_2}.$$

(4.42)

The ratio of the primary and secondary voltages is the same as the *turns ratio*; the greater the number of turns ratio, the greater the voltage ratio. When the number of secondary turns is greater than the number of primary turns, the transformer is called a *step-up transformer* since the secondary voltage is greater than the primary voltage; when the number of primary turns is greater than the number of secondary turns, it is a *step-down transformer*. This property is possible when there is no fringing of the flux and no resistance in the coils.

In addition, with no losses in the transformer, power is conserved and power in equals power out, or

$$P_{IN} = V_1 I_1 = P_{OUT} = V_2(-I_2).$$ (4.43)

Combining Eqs. (4.42) and (4.43), we obtain

$$\frac{-I_2}{I_1} = \frac{V_1}{V_2} = \frac{N_1}{N_2}.$$ (4.44)

The current ratio is inversely proportional to the turns ratio. The primary current of a step-down transformer is less than the secondary current; the inverse is true for step-up transformers. The minus signs in Eqs. (4.43) and (4.44) are due to the definition of power that requires current flow in the direction of the voltage drop. Some authors avoid the minus sign by defining the secondary current oppositely.

Equation (4.44) implies that $N_1 I_1 = -N_2 I_2$, the mmf due to the secondary is equal to that of the primary, but oppositely directed. Consequently, the net flux within the core is zero as the current flow in the secondary and the load produces a flux which counteracts that due to the primary current. In an actual transformer, the secondary flux is oppositely directed, but does not completely cancel the primary flux.

The conditions imposed for the development of Eqs. (4.42) and (4.44)—no fringing fields and no losses—define the *ideal transformer* for which the turns ratio equations are valid. Keep in mind that Eqs. (4.42) and (4.44) are valid for AC signals only. By combining these equations, we obtain an expression for the input impedance of an ideal transformer as

$$Z_{IN} = \frac{V_1}{I_1} = \frac{V_2\left(\frac{N_1}{N_2}\right)}{-I_2\frac{N_2}{N_1}} = \left(\frac{N_1}{N_2}\right)^2\left(\frac{V_2}{-I_2}\right) = \left(\frac{N_1}{N_2}\right)^2 Z_L.$$ (4.45)

The input impedance at the primary of a step-down transformer is larger than the load impedance since $N_1/N_2 > 1$; the converse is true for a step-down transformer. The minus sign is due to the definition of I_2 into the transformer, opposite to the direction of positively defined current in the load Z_L.

When orientations of the turns of the coils on the transformer are unknown, the polarity of the mutual inductance term and the corresponding induced voltages are indicated on circuit diagrams by the notation known as the *dot convention*, see Fig. 4.11. Dots are assigned to one terminal of each coil of the transformer. Both currents entering or both currents leaving the dots of the coils indicates that the fluxes add and that the mutual inductance between those two coils is positive. One current entering and the other leaving a dot indicates that the fluxes are oppositely directed and that the mutual inductance between those two coils is negative. The core with the windings is shown in Fig. 4.11(a); the schematic symbol for the transformer is shown in Fig. 4.11(b).

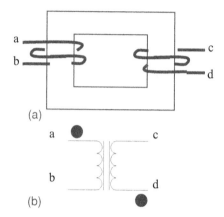

FIGURE 4.11: The dot convention: (a) core with windings and (b) schematic symbol.

When the core and windings are visible, the dots can be determined by inspection. Assign a dot to one terminal of the primary; determine the direction of flux within the core when current flows into this terminal. The dotted terminal of the secondary is that terminal into which current must flow to produce a flux in the same direction. When the coils cannot be seen, the dots can be determined by the application of an AC voltage to the primary coil and the observation of both the primary and secondary voltages. Arbitrarily assign a dot to the positive terminal of the primary coil. The terminal of the secondary that is in-phase with the primary voltage is also dotted. An alternate view of the dot convention is to note that current flowing into the dotted terminal of one winding induces an open-circuited voltage in the other winding which is positive at its dotted terminal.

Example 4.5-1.   An ideal transformer has $N_1 = 100$ and $N_2 = 200$ turns on the primary and secondary coils, respectively. Calculate the voltage and current ratios and the input impedance when a load of 100 Ω is attached to the secondary. The voltage ratio is proportional to the turns ratio as $V_2/V_1 = N_2/N_1 = 200/100 = 2$; the secondary voltage is twice the primary voltage. The current ratio is given as $I_2/I_1 = -N_1/N_2 = -0.5$; the secondary current is half the primary current. The input impedance is given as $Z_{IN} = (N_1/N_2)^2 Z_L = 100/(2)^2 = 25 \Omega$.

Example 4.5-2.   Calculate the input impedance of an ideal transformer with a capacitor as the load. The input impedance is given by $Z_{IN} = (N_1/N_2)^2 Z_L = (1/j\omega C)(N_1/N_2)^2 =$

$1/j\omega[C(N_2/N_1)^2]$. For a step-up transformer, the input "sees" a larger capacitor; for a step-down transformer "sees" a smaller capacitor.

Example 4.5-3.   Choose a transformer to deliver the maximum power from a source with a Thévenin impedance of 100 Ω to a speaker with an impedance of 4 Ω. For maximum power transfer, the source must "see" an impedance of 100 Ω. This is possible if the transformer has a turns ratio described by Eq. (4.45) as $\frac{N_1}{N_2} = \left(\frac{Z_{IN}}{Z_L}\right)^{1/2} = \left(\frac{100}{4}\right)^{1/2} = 5$. Ideal transformers conveniently "alter" impedance values without any losses.

Example 4.5-4.   Determine the location of dots for the circuit model of the ideal transformer of Fig. 4.12(a). Assign a dot to the top terminal of the primary; a current into this terminal establishes a flux in the CW direction within the core. Current flowing into the upper terminal of the primary produces a CW flux, also. Consequently, this terminal is the dotted terminal of the secondary.

4.6   MAGNETIC FORCES

The presence of electric fields produce forces at material boundaries. A similar phenomenon exists in magnetic fields as well. Consider the magnetic field in the gap of a magnetic core as shown in Fig. 4.13.

For simplicity, assume that the air gap is so small that the effects of fringing fields at the edges is negligible. The system is defined as the air gap. The energy contained within the system is given by Eq. 3.48

$$W_m = \frac{B^2}{2\mu_o} A l_g. \qquad (4.46)$$

For constant flux density, the change of system energy as the air gap is increased by an incremental amount δl_g is expressed as

$$\delta W_m = \frac{B^2}{2\mu_o} A \delta l_g. \qquad (4.47)$$

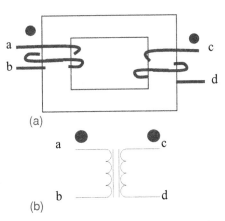

(a)

(b)

FIGURE 4.12:   Dot convention problem: (a) actual transformer and (b) circuit model.

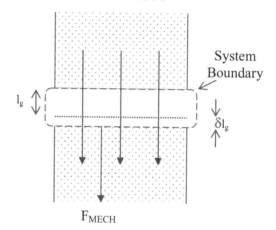

FIGURE 4.13: Virtual work configuration for air gap in magnetic core.

This is accompanied by mechanical work done according to

$$\delta W_{\text{MECH}} = F_{\text{MECH}} \delta l_g. \tag{4.48}$$

From Eqs. (4.47) and (4.48), we obtain the mechanical force required to increase the gap as

$$F_{\text{MECH}} = \frac{B^2}{2\mu_o} A. \tag{4.49}$$

As with the electric fields, the magnetic force is oppositely directed to the mechanical force or

$$F_m = -\frac{B^2}{2\mu_o} A \tag{4.50}$$

and the magnetic pressure on the interface is

$$p_m = -\frac{B^2}{2\mu_o} = -\frac{\mu_o H^2}{2}. \tag{4.51}$$

The minus sign indicates that the magnetic force tends to pull the core faces together reducing the air gap. This is consistent with our earlier observations that nature tries to achieve a state that minimizes the energy of the system. Note that the energy in the gap is reduced as the gap is decreased. The RH portion of Eq. (4.51) is valid for linear material.

Similarly, currents in the presence of a magnetic field are subject to a force, albeit more complicated in nature. To simplify calculations, let's determine the forces between two, identical, parallel sheets of uniform current density, $\mathbf{K}$, separated by a small distance h. The sheets are w deep and L long with $\mathbf{K}$ supplied by a current source of magnitude $I = |\mathbf{K}|w$. The configuration is shown in Fig. 4.14. The current source is shown connected to the sheets of

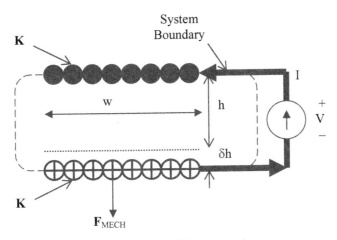

FIGURE 4.14: Virtual work configuration for parallel current sheets.

current, but of course the currents are applied at the ends (infinitely far into or out of the plane of the paper) rather than on the sides. We will use the principle of virtual work to calculate the magnetic forces much as we did in Section 2.12 for calculating electrical forces.

Since the spacing between the currents is small compared to the dimensions of the currents, i.e., $h \ll w$, L, we will assume that the magnetic fields can be approximated by infinite sheets of currents. From Ampere's law, the field between the currents is given by $\mathbf{H} = K\mathbf{a}_X$ A/m while the field outside the currents is zero.

The magnetic energy density is expressed as $w_m = \mu_o |H|^2 / 2$ J/m^3. The total magnetic energy of the system exists only in the region between the currents and is equal to $w_m h t L$ or

$$W_m = \frac{\mu_o K^2 h t L}{2}. \tag{4.52}$$

An incremental increase in the separation of the current sheets results in an increase in magnetic energy between the two currents given by

$$\delta W_m = \frac{\mu_o K^2 w L}{2} \delta h. \tag{4.53}$$

The magnetic energy stored between the two current sheets increases as the sheets are separated because volume occupied by the magnetic fields increases while the energy density is unchanged. An external mechanical force pulls the two current sheets apart, supplying energy to the system according to

$$\delta W_{\text{MECH}} = F_{\text{MECH}} \delta h. \tag{4.54}$$

During the time interval, while the flux between the two sheets is increasing due to the incremental separation, a voltage is generated across the two sheets since the flux linked by the current is increasing. Lenz's law predicts the polarity of the voltage is to oppose the increase in flux, i.e., the voltage drop is in the direction of the current flow. Consequently, the external source supplies energy to the system of

$$\delta W_{\text{EXTSRC}} = IV\delta t = I\left(\frac{\delta\Psi_m}{\delta t}\right)\delta t = I\delta\Psi_m$$

$$= (Kw)(\mu_o KL)\delta h = \mu_o K^2 wL\delta h. \tag{4.55}$$

The energy gained within the system boundary must equal the energy supplied from outside as

$$\delta W_m = \frac{\mu_o K^2 wL}{2}\delta h = \delta W_{\text{MECH}} + \delta W_{\text{EXTSRC}}$$

$$= F_{\text{MECH}}\delta h + \mu_o K^2 wL\delta h. \tag{4.56}$$

Solving for the magnetic force, F_m, against which the mechanical force, F_{MECH}, acts to oppose, we obtain

$$F_m = -F_{\text{MECH}} = \frac{\mu_o K^2 wL}{2}. \tag{4.57}$$

The magnetic force tends to push apart the two oppositely directed current sheets; the magnetic force tends to increase the magnetic flux within the system by increasing the volume between the current sheets. When the two current sheets are in the same direction, the flux between the sheets is zero and the region outside of the two sheets has a nonzero flux. To increase the amount of flux within the system, the magnetic force tends to push the two sheets together. A physical interpretation of this effect is that the *magnetic forces tend to push current carrying elements into regions of lesser magnetic energy density.*

Closed currents in the presence of a magnetic field also experience a torque that tends to align the current so as to maximize the surface perpendicular to the flux. As before, the process tends to maximize the flux through the current carrying element.

The pressure at a boundary can be expressed in terms of energy density. By dividing the force of Eq. (4.57) by the area of the current sheets, we obtain

$$p_m = \frac{\mu_o K^2}{2} = \frac{\mu_o H^2}{2} = w_m \tag{4.58}$$

as the pressure on the current sheet into the region of zero magnetic energy density. A generalization of this equation expresses the pressure as the difference between energy densities at a boundary. In the case of the identically directed current sheets, the field between the sheets has a magnetic energy density of $w_m = \mu_o H^2/2$, the energy density outside the sheets is zero.

The magnetic pressure normal to the boundary between two magnetic materials is expressed as

$$p_m = \frac{(\mu_2 - \mu_1)\left(H_T^2 + \dfrac{B_N^2}{\mu_1 \mu_2}\right)}{2} \tag{4.59}$$

where the pressure is from region 2 to region 1, the higher energy density into the lower. These results are similar to Eqs. (2.72) and (2.74) for the electric forces on a dielectric interface.

Isolated elements of current are not observable; they always form a closed loop of current. Nevertheless, the force on a differential current element as shown in Fig. 4.15 can be determined from the form

$$d\mathbf{F}_m = \mathbf{I}dl \times \mathbf{B}. \tag{4.60}$$

The total force on the closed loop is the integral of the forces on all of the current elements. In a uniform magnetic field, this results in no net force on the loop, but there may be a net torque on the loop.

Example 4.6-1. Calculate the force per unit length on parallel filamentary currents, I_1 and I_2, shown in Fig. 4.15, as a function of their separation. From Ampere's law, the magnetic field of a filament is expressed as $\mathbf{B}_1 = B_{1\phi}\mathbf{a}_\phi = \frac{\mu I_1}{2\pi\rho}\mathbf{a}_\phi$ for current in the z-direction. Equation (4.60) shows that the magnetic force per unit length on the second filament is given by $\mathbf{F}_m = \mathbf{I}_2 \times \mathbf{B}_1 = I_2\mathbf{a}_Z \times \frac{\mu I_1}{2\pi\rho}\mathbf{a}_\phi = -\frac{\mu I_1 I_2}{2\pi\rho}\mathbf{a}_\rho$. The force varies inversely with the separation between the filaments, ρ, and as the product of the current magnitudes, $I_1 I_2$. The minus sign indicates that the

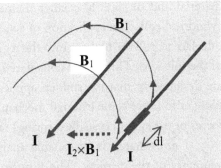

FIGURE 4.15: Magnetic forces on currents.

force tends to pull the two filaments together when the currents are in the same direction. The force tends to push the filaments apart when the currents are oppositely directed as in the case of parallel current sheets.

Example 4.6-2. Calculate the forces on the pole pieces of the core of Example 4.3-1. The gap has a length of 1 mm, a cross-sectional area of 0.0004 m², and a flux of $\Psi_m = 40.2$ µwb. The flux density in both the core and the air gap is given by $B = \Psi_m/A = 0.1$ wb/m² and is normal to the interface. Equation (4.59) gives the pressure on the interface as

$$p_m = \frac{(\mu_2 - \mu_1)}{2}\frac{B_{N^2}}{\mu_1\mu_2} = \frac{\left(\dfrac{1}{\mu_o} - \dfrac{1}{10^3\mu_o}\right)}{2}(.1)^2 = \frac{(1 - 0.001)(0.01)}{2(4\pi \times 10^{-7})} = 4015 \text{ N/m}^2.$$

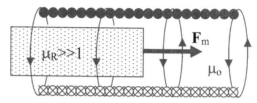

FIGURE 4.16: Cylindrical solenoid.

This leads to a total force on the interface of $F_m = p_m A = 1.61$ N pulling the two pole pieces together.

4.7 SOLENOIDS

The force on the interface between two dissimilar magnetic materials is utilized by magnetic elements known as solenoids, see Fig. 4.16. A typical solenoid is constructed of a cylindrical coil that establishes a nearly-uniform, axial magnetic field, H_Z. An *actuator* or *plunger* of magnetic material and of slightly smaller diameter than the inside of the coil is located within the cylindrical coil. The application of a sufficiently large current causes a force at the plunger–air interface to pull the plunger into the region with less flux, i.e., further into the coil along the axis of the solenoid. This can be considered as a very simple "translational motor." Many solenoids are designed so that the sudden application of current causes a quick, forceful motion of the plunger to move some external mechanical element. When the current ceases, the mechanical forces of the system (usually a spring) return the plunger to its original position.

Unfortunately, a reasonable analysis of the forces on a solenoid is complicated by the geometry of the element. The flux path through the plunger and the core is relatively easy to determine. But the rest of the path by which the flux closes is much more complicated, extending from the open end of the solenoid to the extended plunger on the other end. The path length and cross section are not defined easily since the flux is not confined by a flux guide. Since our mathematical skills are not sufficient to readily solve for the fields in air, we must be content with a previous, more qualitative view. The principle of virtual work describes the nature of the magnetic forces as acting to pull the plunger into the solenoid coil. The magnitude of these forces will be left to a more advanced course.

4.8 RELAYS

Another common magnetic element which converts magnetic energy to motion is the electromechanical relay, see Fig. 4.17. Relays are considered a form of magnetic core with a small air gap and with a portion of the core moveable. Application of a sufficiently large current (*pull-in current*) in the relay coil energizes the core and the resulting magnetic force tends to

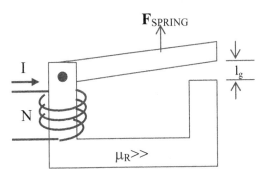

FIGURE 4.17: Electromechanical relay.

pull the moveable actuator toward the fixed portion of the core, completely closing the air gap. Electrical contacts are attached to the actuator. They open or close some electrical circuit with the motion of the actuator. With the interruption of the current, the actuator is returned to its original position by a mechanical spring.

The behavior of a relay is most simply approximated by assuming that the mmf drop within the magnetic material (the relay frame and actuator) is negligible as described in Section 4.3. Therefore, all of the mmf drop occurs across the air gap. The pressure on the actuator is calculated by Eq. (4.59) as

$$p_m = (\mu_2 - \mu_1)\frac{B_N^2}{2\mu_1\mu_2} = \left(1 - \frac{1}{\mu_R}\right)\frac{B_N^2}{2\mu_o} \approx \frac{B_N^2}{2\mu_o} = \frac{\left(\frac{\mu_o NI}{\ell_g}\right)^2}{2\mu_o} = \frac{\mu_o(NI)^2}{2\ell_g^2} \quad (4.61)$$

and the force which pulls the actuator to close the air gap is given by $F_m = p_m A_g = (NI/l_g)^2(\mu_o A_g/2)$. This force must work against the spring that holds the actuator in the open position. The magnetic force is increased by increasing the number of turns, the current, or the air gap area or by reducing the air gap length. The greater the magnetic force, the faster the relay closes. Note that this is the force as the actuator starts to move; the force increases as the actuator moves to decrease the air gap length.

4.9   FARADAY'S LAW REVISITED

As you recall from Chapter 3, the voltage induced in a closed PEC loop is equal to the negative rate of change of the flux enclosed by the loop. In mathematical terms, Faraday's law is expressed by Eq. (3.24) as

$$V_{\text{INDUCED}} = -\frac{d\Psi_m}{dt} = -\frac{d}{dt}\left(\iint_S \mathbf{B} \cdot \mathbf{ds}\right). \quad (4.62)$$

This form is adequate for stationary loops, but it must be modified when the loop is moving. Let **v** represent the velocity of points on the PEC loop. With a bit of advanced vector mathematics, the integral can be decomposed into two parts which are written as

$$V_{\text{INDUCED}} = -\frac{d}{dt} \iint_S \mathbf{B} \cdot \mathbf{ds}$$

$$= -\iint_S \frac{\partial \mathbf{B}}{\partial t} \cdot \mathbf{ds} + \oint_L \mathbf{v} \times \mathbf{B} \cdot \mathbf{dl}. \qquad (4.63)$$

The closed path L is any path that encloses the surface S and coincides with PEC wire. These two forms indicate two different mechanisms for generating induced voltage. The first mechanism is due to time-varying magnetic flux density, i.e., $\mathbf{B} = \mathbf{B}(t)$ and a stationary loop. Though the loop is fixed, the time-varying flux induces a voltage. This is the nature of the magnetic field within wire loops that are fixed in space, e.g., coils, transformers, solenoids, and relays. This mechanism is frequently called *transformer emf*. Chapters 3 and 4 have concentrated upon these devices. The second mechanism is due to changes in the surface S due to motion of the wire PEC loop L. Even with a static magnetic field, the magnetic flux enclosed by the wire loop changes due to motion of the PEC loop. Motion of the loop L alters the flux within the loop and induces a voltage. Motion of L includes rotation or change of size and shape. This process is known as *motional emf*. This mechanism for induced voltage is the basis for electric motors and generators. Many applications involve both transformer and motional emf.

The general form of Faraday's law is rich with subtleties. Therefore, it is often misunderstood and misapplied. In this section, Faraday's law examined by way of several of its simpler applications.* More complicated applications await you in advanced electromagnetics courses.

For the first, several applications consider the situation shown in Fig. 4.18. Two parallel PEC rails lie W distance apart. Both rails are parallel to the x-axis. The rails are connected at the left end by a PEC that lies along the y-axis. A sliding PEC bar at the right end is parallel to the y-axis and slides in the x-direction with velocity **v** in the x-direction. The rails, left end connection, and sliding bar are all in electrical contact and make a closed loop (circuit). A ideal voltmeter of negligibly small dimensions is connected in the near rail (along the x-axis). A magnetic field with uniform and constant flux density **B** in the z-direction is present throughout the region of the rails.

Application 1: The induced voltage V in the closed loop (as measured by the ideal voltmeter) can be calculated via Eq. (4.62) by direct consideration of the flux enclosed by the PEC loop.

---

* These examples were suggested by Prof. Frank Acker, ECE Department, Rose-Hulman Institute of Technology.

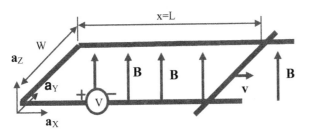

FIGURE 4.18: Sliding bar on rails.

A more general form of the flux enclosed by an N-turn loop uses the concept of flux linkages, λ, from power engineering as

$$V_{\text{INDUCED}} = -\frac{d\lambda}{dt} = -\frac{d(N\Psi_m)}{dt} = -N\frac{d\Psi_m}{dt}. \qquad (4.64)$$

As the bar slides in the x-direction, the increasing area of the circuit encloses more flux. Thus, there is a $d\Psi_m/dt$ and the voltmeter indicates a nonzero value. By Lenz's law, the polarity of the induced voltage will tend to set up a current that tries to keep the enclosed flux a constant. This flux must be downward to oppose the increasing upward flux; by the RH rule, the current must flow in the CCW direction. Therefore, the voltmeter will read a negative value. Mathematically, this is expressed as follows. For the system shown in Fig. 4.18, the flux enclosed by the PEC loop is $\Psi_m = \iint_S \mathbf{B} \cdot \mathbf{ds} = BWx$ where $x = L$. The number of turns $N = 1$. Using Eq. (4.64), we calculate the induced voltage as

$$V_{\text{INDUCED}} = -\frac{d\lambda}{dt} = -1\frac{d\Psi_m}{dt} = -BW\frac{dx}{dt} = -vBW. \qquad (4.65)$$

Application 2: The induced voltage of the rail system can also be calculated by means of the two-term form of Faraday's law expressed by Eq. (4.63). As a first step, a stick figure is attached to show the location of the observer as shown in Fig. 4.19, the location of the stationary observer. Note that the two rails and the left end connection are at rest relative to the observer.

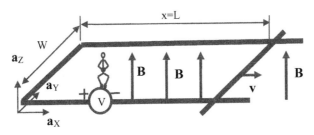

FIGURE 4.19: Stationary observer in rail system.

The induced voltage is calculated by means of Eq. (4.63)

$$V_{\text{INDUCED}} = -\iint_S \frac{\partial \mathbf{B}}{\partial t} \cdot d\mathbf{s} + \oint_L \mathbf{v} \times \mathbf{B} \cdot d\mathbf{l} \qquad (4.66)$$

Since the observer sees a constant **B**-field within the loop, the derivative of **B** with respect to time is zero so that the first term due to transformer emf vanishes, i.e., $-\iint_S \frac{\partial \mathbf{B}}{\partial t} \cdot d\mathbf{s} = 0$. The closed path for the second term coincides with the PEC conductors, i.e., both rails, the left end connections, and the sliding bar. Since both rails and the left end connection are stationary relative to the observer, $\mathbf{v} = 0$ on all segments of the path except the sliding bar. On this segment, $\mathbf{v} = v\mathbf{a}_X$. The direction of integration is CCW as required by the left terminal of the voltmeter defined as positive. Therefore, the induced voltage is calculated as

$$V_{\text{INDUCED}} = \oint_L \mathbf{v} \times \mathbf{B} \cdot d\mathbf{l}$$

$$= \int_{y=0}^{W} v\mathbf{a}_X \times B\mathbf{a}_Z \cdot \mathbf{a}_Y dy$$

$$= -vWB. \qquad (4.67)$$

Recall that the direction of the path **dl** is chosen by the RH rule with the thumb pointing in the direction of **ds** of the enclosed surface. Therefore, the path of the line integral is CCW since **ds** is aligned with the positive direction of flux as upward through the loop. Note that the negative sign comes from the vector operations inside the integral. As expected, the result is the same as the previous calculation. However, it is mathematically more straightforward and it does not rely on Lenz's law to establish the polarity of the induced voltage.

Application 3: Another method of calculating the voltage induced in the sliding rail system is to consider the observer to be on the sliding bar as shown in Fig. 4.20.

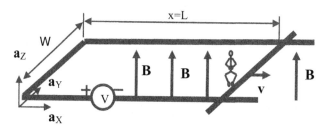

FIGURE 4.20: Moving observer in rail system.

As before, the observer sees a constant **B**-field so that

$$-\iint_S \frac{\partial \mathbf{B}}{\partial t} \cdot \mathbf{ds} = 0. \tag{4.68}$$

On the other hand, the sliding bar is at rest with respect to the observer and the rails and left end connection appear to be moving in the $-x$-direction, i.e., $\mathbf{v} = -v\mathbf{a}_X$. Starting at the origin and evaluating the line integral along the path composed of the rails, end connection, and bar yields

$$\oint_L \mathbf{v} \times \mathbf{B} \cdot \mathbf{ds} = \int_{\substack{\text{NEAR} \\ \text{BAR}}} v B \mathbf{a}_Y \cdot \mathbf{a}_X dx$$

$$+ \int_{\substack{\text{SLIDING} \\ \text{BAR}}} 0 \cdot \mathbf{a}_Y dy + \int_{\substack{\text{FAR} \\ \text{BAR}}} v B \mathbf{a}_Y \cdot \mathbf{a}_X dx$$

$$+ \int_{\substack{\text{FIXED} \\ \text{BAR}}} v B \mathbf{a}_Y \cdot \mathbf{a}_Y dy = \int_{y=W}^{0} v B dy = -v W B. \tag{4.69}$$

Application 4: The rail system is more complicated when the magnetic flux density is time varying, see Fig. 4.21. With the observer located on the fixed voltmeter, both terms of Faraday's law are required to calculate the induced voltage. For $\mathbf{B} = \mathbf{a}_Z B_{MAX} \cos \omega t$, the surface integral is calculated as

$$-\iint_S \frac{\partial \mathbf{B}}{\partial t} \cdot \mathbf{ds} = -\int_{x'=0}^{x} \int_{y=0}^{W} B_{MAX} \frac{\partial \cos \omega t}{\partial t} dy dx$$

$$= \int_{x'=0}^{x} \int_{y=0}^{W} \omega B_{MAX} \sin \omega t \, dy dx$$

$$= x W \omega B_{MAX} \sin \omega t. \tag{4.70}$$

Defining the position of the sliding bar to be given by $x = L = L_o + vt$, we can write the transformer emf as

$$V_{TRANS} = (L_o + vt) \omega W B_{MAX} \sin \omega t. \tag{4.71}$$

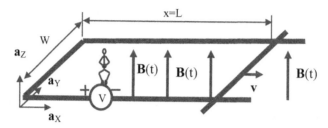

FIGURE 4.21: Time-varying magnetic field in rail system.

The motional emf is similar to Eq. (4.67) where B is time-varying as

$$V_{\text{MOTIONAL}} = -vWB_{\text{MAX}} \cos \omega t. \tag{4.72}$$

The total voltage is expressed as

$$V_{\text{INDUCED}} = (L_o + vt)\omega WB_{\text{MAX}} \sin \omega t$$
$$- vWB_{\text{MAX}} \cos \omega t. \tag{4.73}$$

Application 5: Induced voltage can be generated by rotary motion as well as by linear motion of the first four applications. The Faraday disk shown in Fig. 4.22 is one example.

The flux density **B** is constant and uniform throughout the region of the disk. The disk is made of a metallic conductor such as copper (as usual, assume it is a PEC) and is rotated in the magnetic field at angular speed ω. The axle of the disk is parallel to the direction of **B**. The radius of the disk is R_2. The radius of the axle is R_1. Sliding brushes make electrical contact at the axle and outer edge of the disk. The sliding brushes are usually made of carbon blocks. The

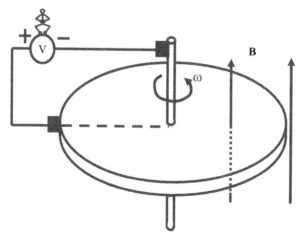

FIGURE 4.22: Faraday disk generator with fixed path.

current drawn by the voltmeter is so small that the voltage drop throughout the carbon brush is negligible.

Define the path and the observer location. The path shown here lies in the voltmeter, its connecting wires, the brushes, the dotted path along the surface of the disk, and a vertical line along the surface of the axle to form a closed loop. No part of the path moves with respect to the observer. Because **B** is constant, the first term of Faraday's law vanishes, i.e., $-\iint_S \frac{\partial \mathbf{B}}{\partial t} \cdot \mathbf{ds} = 0$. For the line integral term, the voltmeter and wires have zero velocity with respect to the observer. Although the dashed line segment does not move, the conductor (the surface of the disk) at the location of the path moves with a velocity that depends upon the radial location as $v = r\omega$ and is perpendicular to **B**. The conductive axial portion of the path also moves with velocity $R_1\omega$ perpendicular to **B**, but along this portion of the path $\mathbf{v} \times \mathbf{B}$ is perpendicular to the **dl** direction of integration on the path. Therefore, the induced voltage is expressed exclusively by the line integral as

$$V_{\text{INDUCED}} = \oint_L \mathbf{v} \times \mathbf{B} \cdot \mathbf{dl}$$

$$= \int_{\text{WIRES}} 0 \cdot \mathbf{dl} + \int_{\text{AXIAL}} v\,B\mathbf{a}_\phi \cdot \mathbf{dl} + \int_{R_1}^{R_2} \omega r\,dr$$

$$= \omega \frac{(R_2^2 - R_1^2)}{2} \xrightarrow[\lim R_1 \to 0]{} \omega \frac{R_2^2}{2}. \tag{4.74}$$

Usually $R_2 \gg R_1$ so that R_1 can be neglected. Note that the polarity of the meter reading can be checked by means of the right-hand rule. An electron within the conductor moves in the CCW direction. According to $\mathbf{F} = q\mathbf{v} \times \mathbf{B}$, it will be "pushed" toward the axial and the negative terminal of the voltmeter. This will create an absence of electrons out the outer edge of the rim and it will be positive.

Application 6: Consider the same system as in Application 5, with the observer in the same location, but with the path as shown in Fig. 4.23. The difference between this and the previous calculation is that now the dotted-line portion of the path is fastened to the disk. One segment of this path on the disk is a radial section r that is "stuck" to the surface of the disk and rotates through ever-increasing angle θ as the disk spins. A second segment of this path is tangential to the edge of the disk and "stuck" to it. This second segment, denoted as $t = R_2\theta$, gets longer and longer as the disk spins. Before we do the math, note that the flux enclosed by the path is increasing due to increasing angle θ so we expect an induced voltage.

As before, because B is constant as seen by the observer, the first term of Faraday's law vanishes $-\iint_S \frac{\partial \mathbf{B}}{\partial t} \cdot \mathbf{ds} = 0$. Therefore, the induced voltage is due to only the radial component

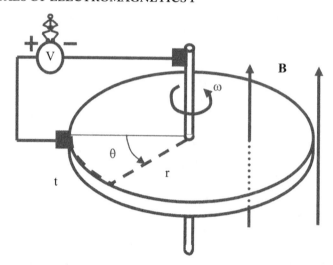

FIGURE 4.23: Faraday disk generator with moving path.

of the path, r. The calculation is identical to that in Application 5

$$V_{\text{INDUCED}} = \oint_L \mathbf{v} \times \mathbf{B} \cdot \mathbf{dl}$$

$$= \int_{\text{WIRES}} 0 \cdot \mathbf{dl} + \int_{\text{AXIAL}} v B \mathbf{a}_\phi \cdot \mathbf{dl} + \int_{R_1}^{R_2} \omega r \, dr$$

$$= \omega \frac{\left(R_2^2 - R_1^2\right)}{2} \xrightarrow[\lim R_1 \to 0]{} \omega \frac{R_2^2}{2}. \qquad (4.75)$$

Note that *v is the velocity of the conductor at the location of the path* that is used in the formula. The velocity of the path or whether the path has a velocity is not considered. Accordingly, the component along the tangential part of the path is zero since $\mathbf{v} \times \mathbf{B}$ is perpendicular to the direction of integration along the tangential path.

Application 7: This application is based on the excitation by a DC primary current of a variable transformer (Variac). The two-term integral form of Faraday's law is especially useful in verifying that transformers do not function for DC currents and voltages. (But, you know that from circuits!)

As shown in Fig. 4.24, a constant current in the primary winding establishes a constant magnetic flux in the iron core. The individual turns of the secondary winding are insulated from the core and each other. The wire used to make the secondary winding is bare (uninsulated) so that contact can be made by means of a sliding carbon brush to the voltmeter. Since the ideal

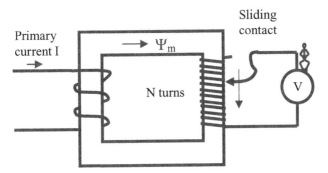

FIGURE 4.24: DC variac.

voltmeter draws no current, the voltmeter and sliding brush cannot affect the flux established by the primary current.

An important question is stated as follows. If the sliding contact is quickly moved from a position (near the top) where many turns are linked to the voltmeter to a new position (near the bottom) where only a few turns are linked to the voltmeter, will the voltmeter sense a transient voltage as the brush slides along the turns? Clearly the number of flux linkages is changing and a naive understanding of Faraday's law in terms of flux linkages might lead us to believe that a voltage will be induced in the voltmeter as the brush slides. However, experiment shows that the voltmeter does not detect a transient voltage. The two-term expression of Faraday's Law easily yields the correct answer.

The closed circuit used to calculate the voltmeter reading includes the voltmeter, the connecting wires, the sliding contact, and those turns below the sliding contact that are connected to the voltmeter leads. The observer is assumed to the standing on the voltmeter. For conceptual simplicity, consider that all of the flux due to the primary current is in the iron core (no leakage flux). Since the flux is constant, the flux density **B** does not vary with time, and the partial derivative of **B** with respect to time is zero. Since there is no flux density outside the core, the **B** in the path integral is zero. Hence, as the brush slides, the voltmeter reads zero as expressed by

$$V_{\text{INDUCED}} = -\iint_S \frac{\partial \mathbf{B}}{\partial t} \cdot \mathbf{ds} + \oint_L \mathbf{v} \times \mathbf{B} \cdot \mathbf{dl} = 0 + 0 = 0. \qquad (4.76)$$

Application 8: You might be suspicious of the above example because the number of turns is switched discontinuously as the contact slides along the turns of wire. Consider the circuit in Fig. 4.25 that has no switching action and is completely continuous. The right leg of the iron core is cylindrical and can freely rotate as the wire is pulled and removed from the core. A slip-ring maintains a continuous connection to the voltmeter at the top end of the winding.

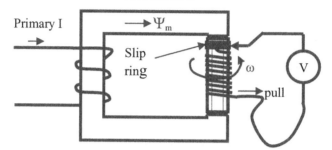

FIGURE 4.25: Unwinding coil.

Again, for conceptual simplicity, consider that the only magnetic field is inside the iron core and the air gaps at the top and bottom of the cylindrical leg of the core. The observer stands on the voltmeter. No electrical conductor is in the magnetic field although they link it.

The wire is pulled from the iron causing the right leg of the core to rotate so that the number of turns of wire that link the flux is changing in a continuous manner. Will the meter detect a voltage as the wire is pulled from the core?

The reasoning is identical with that of the previous example. There is no time variation of magnetic flux density. No part of the conducting path is in the magnetic field. Thus, there is no induced voltage

$$V_{\text{INDUCED}} = -\iint_S \frac{\partial \mathbf{B}}{\partial t} \cdot \mathbf{ds} + \oint_L \mathbf{v} \times \mathbf{B} \cdot \mathbf{dl} = 0 + 0 = 0. \qquad (4.77)$$

These eight applications show the variety of calculation methods for Faraday's law. The transformer emf contributions are fairly straightforward. The magnetic flux is changing with time and induces a voltage in the circuit. The motional emf is a bit more complicated. The mathematics shows that $\mathbf{v} \times \mathbf{B}$ must lie along at least a portion of the path $\mathbf{dl}$. An alternate way of saying this is that *the conductors of the circuit must "cut across" magnetic field lines.* Any motion that doesn't cut flux lines doesn't induce any voltage.

In spite of the intrigue of Faraday's law, we must move on to other topics. Never fear, we will revisit and utilize Faraday's law repeatedly in future chapters.

The topics covered in the first four chapters have focused upon the behavior of electromagnetic field within lumped elements that are small. But, we have not really defined what we mean by "small"? In fact, we mean that the frequency of operation is so low that the wavelength associated with this frequency is very much larger than every dimension of the lumped element. In *Fundamentals of Electromagnetics Volume 2: Quasistatics and Waves*, the second volume of this series, we shall examine the behavior of electromagnetic fields when the frequency of operation is much higher. I hope that you can join me on this journey into the really exciting aspects of electromagnetics!!

Author Biography

Dr. David Voltmer, Emeritus Professor of Electrical and Computer Engineering at Rose-Hulman Institute of Technology, received his formal education at Iowa State University (BS EE), University of Southern California (MS EE), and The Ohio State University (PhD EE). Informally, Dave has learned much from his many students while teaching at Pennsylvania State University and at Rose-Hulman. Dave (aka Dr. EMAG) has taught electromagnetic fields and waves, microwaves, and antennas for nearly four decades. Throughout his teaching career, Dave has focused on improving teaching methods and developing courses to keep pace with technological advancements. Spare moments not spent with his family are occupied by long distance bicycling and clawhammer style banjo. Dave is an ASEE Fellow and an IEEE Life Senior Member.

Printed in the United States
by Baker & Taylor Publisher Services